Lecture Notes in Computer Science   8780

Commenced Publication in 1973
Founding and Former Series Editors:
Gerhard Goos, Juris Hartmanis, and Jan van Leeuwen

Editorial Board

More information about this series at http://www.springer.com/series/8851

Ryszard Kowalczyk · Ngoc Thanh Nguyen (Eds.)

Transactions on Computational Collective Intelligence XVI

 Springer

Editor-in-Chief

Ngoc Thanh Nguyen
Institute of Informatics
Wroclaw University of Technology
Wroclaw
Poland

Co-editor-in-Chief

Ryszard Kowalczyk
Swinburne University of Technology
Hawthorn Vic
Australia

ISSN 0302-9743
ISBN 978-3-662-44870-0
DOI 10.1007/978-3-662-44871-7

ISSN 1611-3349 (electronic)
ISBN 978-3-662-44871-7 (eBook)

Library of Congress Control Number: 2014949397

Springer Heidelberg New York Dordrecht London

Printed on acid-free paper

Springer is part of Springer Science+Business Media (www.springer.com)

Preface

Welcome to the 16th volume of Transactions on Computational Collective Intelligence (TCCI). This volume of TCCI includes seven interesting and original papers selected via a peer-review process. The first paper, entitled "Non-intrusive Repair of Safety and Liveness Violations in Reactive Programs" by Harel et al., shows how, under certain conditions, programs written in the behavioral programming approach can be modified using automatically generated code modules. At the core of the presented approach is the ability of a thread of behavior to prevent the triggering of events from other threads. The proposed repair algorithms apply model checking of safety and liveness properties to the program and transform the counterexamples produced by the model-checker into corrective modules. In the second paper "Designing Adaptive Systems Using Teleo-Reactive Agents," the authors Smith et al. provide a formal definition of adaptivity which is independent of any particular adaptivity mechanism and general enough to use with any specification notation. Then based on this definition, a framework for specifying the behavioral requirements of adaptive agents and MAS in a systematic way is presented. The third paper, "Towards Formal Modelling and Verification of Pervasive Computing Systems" by Liu et al., proposes the use of formal methods to analyse pervasive computing systems. It includes a formal modeling framework to cover the main characteristics of such systems, and identification and specification of the safety requirements as safety and liveness properties. Furthermore, based on the modeling framework, the authors propose an approach of verifying reasoning rules that are used in the middleware for perceiving the environment and making adaptation decisions. Experimental results show the usefulness of the presented approach in exploring system behaviors and revealing system design flaws such as information inconsistency and conflicting reminder services. In the fourth paper "Revisiting Agent-Based Models of Algorithmic Trading Strategies" the authors, Natalia Ponomareva and Anisoara Calinescu, present a method for identifying the most suitable market simulation type based on the specific market model to be investigated. Then the authors propose an extended model of the Bayesian execution strategy for Algorithmic Trading that aims at executing large orders discretely, in order to minimize the orders' impact, while also hiding the traders' intentions. The authors of the fifth paper "Self-explanation in Adaptive Systems based on Runtime Goal-Based Models" by Welsh et al., argue that a self-adaptive system's behavior needs to be explained in terms of satisfaction of its requirements. Since self-adaptive system requirements may themselves be emergent, the authors propose the use of goal-based requirements models at runtime to offer self-explanation of how a system is meeting its requirements. In the sixth paper "A Higher-Order Agent Model with Contextual Planning Management for Ambient Systems" by Chaouche et al., the authors present a concrete software architecture dedicated to ambient intelligence features and requirements. The proposed behavioral model, called higher-order agent captures the evolution of the mental representation of the agent and that of its plan simultaneously. Plan expressions are written and composed using a formal algebraic language, so that plans are built automatically and on the fly, and the updates of sub-plans are realized automatically according to the revising of intentions, hence maintaining the consistency of the agent. The last paper, entitled "An Ontological Consensus Augmented Framework for Collaborative Business Process Formulation" by Hoang et al., presents an ontological approach for forming collaboration of business processes agilely within a BizKB framework that allows to take into account the challenge of dynamically forming collaborative business processes. Moreover, the authors use the consensus theory in distributed knowledge processing in the presented method for service discovery.

TCCI is a peer-reviewed and authoritative journal dealing with the working potential of computational collective intelligence (CCI) methodologies and applications, as well as emerging issues of interest to academics and practitioners. The research area of CCI has been growing significantly in recent years and we are very thankful to everyone within the CCI research community who has supported the Transactions on Computational Collective Intelligence and its affiliated events including the International Conferences on Computational Collective Intelligence (ICCCI).

We would like to thank all the authors, Editorial Board members, and the reviewers for their contributions to TCCI. We express our sincere gratitude to all of them. Finally, we would also like to express our thanks to the LNCS editorial staff of Springer with Alfred Hofmann, who support the TCCI journal.

July 2014

Ryszard Kowalczyk
Ngoc Thanh Nguyen

Transactions on Computational Collective Intelligence

This Springer journal focuses on research on the applications of computer based methods of computational collective intelligence (CCI) and their applications in a wide range of fields such as the Semantic Web, social networks, and multi-agent systems. It aims to provide a forum for the presentation of scientific research and technological achievements accomplished by the international community.

The topics addressed by this journal include all solutions to real-life problems, for which it is necessary to use CCI technologies to achieve effective results. The emphasis of the papers is on novel and original research and technological advancements. Special features on specific topics are welcome.

Editor-in-Chief

Edward Szczerbicki   University of Newcastle, Australia
Tadeusz Szuba   AGH University of Science and Technology,
                              Poland
Kristinn R. Thorisson   Reykjavik University, Iceland
Gloria Phillips-Wren   Loyola University Maryland, USA
Sławomir Zadrożny   Institute of Research Systems, PAS, Poland
Bernadetta Maleszka   Assistant Editor, Wroclaw University
                              of Technology, Poland

Contents

Non-intrusive Repair of Safety and Liveness Violations in Reactive Programs

David Harel[1], Guy Katz[1]([✉]), Assaf Marron[1], and Gera Weiss[2]

[1] Department of Computer Science and Applied Mathematics,
Weizmann Institute of Science, Rehovot, Israel
{david.harel,assaf.marron,guy.katz}@weizmann.ac.il
[2] Department of Computer Science,
Ben-Gurion University of the Negev, Beer-Sheva, Israel
geraw@cs.bgu.ac.il

Abstract. We show how, under certain conditions, programs written in the *behavioral programming* approach can be modified (e.g., as a result of new requirements or discovered bugs) using automatically-generated code modules. Given a trace of undesired behavior, one can generate a relatively small piece of code, whose execution is interwoven at run time with the rest of the system, and which brings about the desired changes without modifying existing code and without introducing new bugs. At the core of our approach is the ability of a thread of behavior to prevent the triggering of events from other threads. Our repair algorithms apply model checking of safety and liveness properties to the program and transform the counterexamples produced by the model-checker into corrective modules. The work is supported by a proof-of-concept tool, which creates understandable modules that can be further manually managed as part of a process of ongoing incremental system development.

Keywords: Program repair · Verification · Behavioral programming · Model checking · Patching

1 Introduction

Software maintenance is a difficult and error prone task. As errors (bugs) are discovered and requirements are added or changed, developers work hard to modify existing code without introducing new errors. They are often constrained by limited knowledge of possible side-effects, since undocumented interdependencies might have been forgotten or might be known only to different people (usually, the original developers) who are unavailable. Research on automated program repair and, more generally, program synthesis from specifications, aims to address these and related challenges. Such automation may prove particularly valuable for handling failure/bug reports from users who simply press the *"Send to Software Vendor"* button. In such cases, the software engineer cannot discuss with the user the context of the problem, or possible generalizations thereof.

© Springer-Verlag Berlin Heidelberg 2014
R. Kowalczyk and N.T. Nguyen (Eds.): TCCI XVI, LNCS 8780, pp. 1–33, 2014.
DOI: 10.1007/978-3-662-44871-7_1

In this paper we focus on programs written in the *behavioral programming* approach, and our work is centered on the idea of repairing by carefully forbidding existing faulty execution paths. This technique is highly suitable for (a) non-intrusive incremental repair; i.e., large parts of the system are already developed and are not modified by the repair process; (b) methodological integration of the repair process with standard, ongoing development during and after the repair activity; and (c) practical techniques for dealing with the complexity of the use of model-checking when creating local patches in the repair process.

2   Background

Our work is carried out within the *behavioral programming* approach [11,12] — an extension and generalization of scenario-based programming, which was introduced with the language of live sequence charts (*LSC*s) [6,10], and is now implemented also in Java [11] and Erlang [13,25].

A behavioral program consists of independent threads of behavior that are interwoven at run time. Each *behavior thread* (abbr. *b-thread*) specifies events and event sequences which, from its own point of view must, may, or must not occur. As shown in Fig. 1, the infrastructure synchronizes and interweaves all behaviors, selecting events that constitute integrated system behavior without requiring direct communication between b-threads. Specifically, all b-threads declare events that should be considered for triggering (called *requested events*) and events whose triggering they forbid (*block*), and then synchronize. An event selection mechanism then triggers one event that is requested and not blocked, and resumes all b-threads that requested the event. B-threads can also declare events that they simply "listen-out for", and they too are resumed when these waited-for events occur.

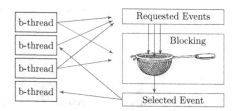

Fig. 1. Behavioral programming execution cycle: all b-threads synchronize, declaring requested and blocked events; a requested event that is not blocked is selected and b-threads waiting for it are resumed.

This facilitates incremental non-intrusive development as outlined in the example of Fig. 2.

More detailed examples showing the power of incremental modularity in behavioral programming appear in [11,13]. Briefly, in a program we wrote for playing Tic-Tac-Toe [11], each game-rule is implemented in a dedicated b-thread;

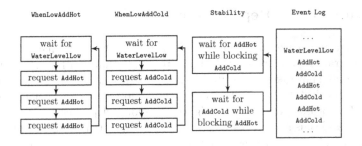

Fig. 2. Incremental development of a system for controlling water level in a tank with hot and cold water sources. The b-thread `WhenLowAddHot` repeatedly waits for `WaterLevelLow` events and requests three times the event `AddHot`. `WhenLowAddCold` performs a similar action with the event `AddCold`, reflecting a separate requirement, which was introduced when adding three water quantities for every sensor reading proved to be insufficient. When `WhenLowAddHot` and `WhenLowAddCold` run simultaneously, with the first at a higher priority, the runs will include three consecutive `AddHot` events followed by three `AddCold` events. A new requirement is then introduced, to the effect that water temperature should be kept stable. We add the b-thread `Stability`, to interleave `AddHot` and `AddCold` events. For details about how sensor and actuator b-threads interact with the physical environment (sensors, valves) without suspending the entire system see [13].

e.g. *"block X moves when it is O's turn"* or *"block marking of already-marked squares"*. Similarly, player-strategy modules are oblivious of other strategies; e.g., *"wait for two X marks in the same line, and then request marking O in that line"*. A similar technique can be used to control a robot performing simultaneous missions, such as vehicle operation and route management. In stabilizing a quadrotor — an unmanned flying vehicle with four rotors — each of four b-threads in our program controls a particular orientation angle, or the quadrotor's altitude, solely by changing rotor speeds; see [13].

Each b-thread repeatedly requests and blocks events representing possible increases or decreases of rotor RPM, which could contribute to its own goal. The triggering of an event that is requested by one or more b-threads and blocked by none allows at least one b-thread to progress. Affected b-threads can then recalculate their declarations of requested and blocked events, and the process repeats.

In [8] and [15], model-checking and planning algorithms (respectively) are applied to play-out, the method for executing LSCs. These smart play-out techniques control the choice of the event to be triggered, such that, within the next superstep (i.e., prior to the next event driven by the environment), the specification is not violated by the program (if this is possible). In [9], a proof-of-concept model checker verifies behavioral Java programs "in vivo" - without first translating them into a model-checker-specific language. It is further shown in [9] how, when a problem is detected, the programmer can develop and add a b-thread that repairs the program by refining the behavior without modifying existing code.

3   Outline of the Repair Approach

In the present paper we utilize the model checker of [9] to automate elements of manual program-repair processes, using a principle that can be summarized as "taking the road not taken". For illustration, assume that a system was tested, or even model-checked, to satisfy its specification, and a new requirement was then introduced, or a bug reported, highlighting a required property not previously articulated, and thus neither tested nor model-checked. Our method calls for first adding the new property to the specification. We then model-check the program to find distinct violating runs. In the case of violated *safety* properties ("bad things never happen"), for each such run we add a special b-thread that waits for the sequence of all events in the run, up to the last one requested by the program (rather than by the environment). The repair b-thread then blocks this event. Some other pending requests might then be triggered. Violated *liveness* properties ("good things eventually happen") are handled similarly: when the system is traversing a loop in which "good things do not happen", the repair b-thread applies blocking to steer the run in another direction. In the liveness case blocking is only performed with some small probability, thus injecting bias towards certain desirable execution paths without forbidding other paths which are also permitted.

For example, consider a faulty game-strategy b-thread, whose event request leads to a loss. When this event is blocked, another b-thread, perhaps one that requests a set of default moves, comes into play (so to speak), offering an alternative. The elimination process continues until "the right" default move is the choice at that state. The new corrective wait-and-block behavior is non-intrusive, in that its implementation does not require changing the existing program code.

We refer to such a repair b-thread as *a patch*, and to the process as *patching*, or simply, *repairing*. We hope that combined with the behavioral-programming principles, our approach will help make the concept of patching seem less a "necessary evil" and more a useful, mainstream software maintenance practice.

As the full repair algorithm may not scale up to large programs due to the state explosion problem, we also discuss the case where patching can be limited to a bounded "neighborhood" of a specific operation scenario; for example, when we are provided with a bug report sent from a user.

We formally prove correctness and analyze the method, characterize the programs on which it can be used, and exemplify its usage with our proof-of-concept tool.

The rest of this paper is organized as follows. Basic definitions of behavioral programs and their model-checking are given in Sects. 4 and 5, respectively. The repair of safety violations of loopless programs is discussed in Sect. 6, followed by a repair algorithm for safety violations in general programs in Sect. 7. Next, in Sect. 8 we extend the algorithm to also handle liveness violations. A method for handling large programs, called limited-depth patching, is described in Sect. 9. Related work is discussed in Sect. 10, and we conclude with Sect. 11.

4   Definitions

While behavioral programming is geared towards natural and intuitive development using almost any programming language, its underlying infrastructure can be conveniently described and analyzed in terms of transition systems.

4.1   The Behavioral Programming Computational Model

The definitions below follow [9,13] and were modified to include the notion of a b-thread tagging states of the system as having certain properties, commonly termed *atomic propositions (AP)* [3]. Recall that a *deterministic labeled transition system* is a 6-tuple $\langle S, E, \rightarrow, init, AP, L \rangle$, where S is a set of states, E is a set of events, $\rightarrow$ is a (possibly partial) function from $S \times E$ to S, $init \in S$ is the initial state, AP is a set of *atomic propositions*, and $L : S \rightarrow 2^{AP}$ is a labeling function. The *runs* of a transition system are sequences of the form $s_0 \xrightarrow{e_1} s_1 \xrightarrow{e_2} \cdots \xrightarrow{e_i} s_i \cdots$, where $s_0 = init$, and for all $i = 1, 2, \cdots$, $s_i \in S$, $e_i \in E$, and the function $\rightarrow$ maps the pair $\langle s_{i-1}, e_i \rangle$ to s_i, written $s_{i-1} \xrightarrow{e_i} s_i$. We say that $\langle S, E, \rightarrow, init \rangle$ is total if the function $\rightarrow$ is total.

Behavior threads are modeled as transition systems, with S, E, and AP finite, and the states being associated with event sets:

Definition 1. *A* behavior thread *(abbr. b-thread) is a tuple $\langle S, E, \rightarrow, init, AP, L, R, B \rangle$, where $\langle S, E, \rightarrow, init, AP, L \rangle$ forms a deterministic total labeled transition system, $R: S \rightarrow 2^E$ associates a state with the set of events requested by the b-thread when in it, and $B: S \rightarrow 2^E$ associates a state with the set of events blocked by the b-thread when in it.*

Definition 2. *The runs of a set of b-threads $\{\langle S_i, E_i, \rightarrow_i, init_i, AP_i, L_i, R_i, B_i \rangle\}_{i=1}^n$ are the runs of the labeled transition system $\langle S, E, \rightarrow, init, AP, L \rangle$, where $S = S_1 \times \cdots \times S_n$, $E = \bigcup_{i=1}^n E_i$, $init = \langle init_1, \ldots, init_n \rangle$, and $\rightarrow$ includes a transition $\langle s_1, \ldots, s_n \rangle \xrightarrow{e} \langle s_1', \ldots, s_n' \rangle$ if and only if*

$$\underbrace{e \in \bigcup_{i=1}^n R_i(s_i)}_{e \text{ is requested}} \qquad \bigwedge \qquad \underbrace{e \notin \bigcup_{i=1}^n B_i(s_i)}_{e \text{ is not blocked}}$$

and

$$\bigwedge_{i=1}^n \Big(\underbrace{(e \in E_i \implies s_i \xrightarrow{e}_i s_i')}_{\substack{\text{affected b-threads} \\ \text{move}}} \wedge \underbrace{(e \notin E_i \implies s_i = s_i')}_{\substack{\text{unaffected b-threads} \\ \text{don't move}}} \Big).$$

We set $AP = \bigcup_{i=1}^n AP_i$ and, for $(s_1, \ldots, s_n) \in S_1 \times \ldots \times S_n$, we define:

$$L(s_1, \ldots, s_n) = L_1(s_1) \cup \ldots \cup L_n(s_n).$$

Note that when implemented in a standard programming language, we assume that b-threads do not share data, and rely solely on events for input and output. This results in the abstraction that a behavior thread is "in a state" only when synchronized with others, and that the state transition caused by executing program instructions between synchronization points is atomic.

Observe that while each b-thread is deterministic in its reaction to events, Definition 2 does not specify how events are selected, and thus there may be more than one run for a given set of b-threads. There could be multiple ways to select events and runs, including ones that are random, planned, or priority-based. The default behavioral execution infrastructure of LSC (in the *Play-Engine* and *PlayGo* tools), the Java package (*BPJ*) and the Erlang module (*bp*) executes a set of b-threads based on priorities. That is, in each state of the composite system, the first event that is requested and is not blocked is selected for triggering.

Definition 3. *For the transition system T, defined in Definition 2, a (deterministic) event selection mechanism is a function $f: S \to E$, such that for each $s \in S$ there exists a transition $s \xrightarrow{f(s)} s'$ of T.*

Behavioral programming is designed particularly for the development of *reactive systems* [14], and in this context it is critical to distinguish between environment behavior and program behavior.

Definition 4. *A reactive behavioral program is a set of b-threads, an event selection mechanism, and a partition of the events of the b-threads into external events representing uncontrollable occurrences coming from the environment, and internal events completely controlled by the program.*

We denote the set of external events by E_{env}, and the set of internal events by E_{prog}. By convention, the patches we present in this work may block only the triggering of events in E_{prog} and may not block events in E_{env}.

4.2 Specifications

We now introduce definitions that assist in the discussion of desired and undesired runs of behavioral programs.

Definition 5. *For a set of b-threads P and a run $\rho = (e_1, e_2, \ldots,)$, such that the execution corresponding to ρ is $s_{init} \xrightarrow{e_1} s_1 \xrightarrow{e_2} s_2 \ldots$, we define $APtrace(\rho) = L(s_{init})L(s_1)L(s_2) \ldots$ and define the set of all traces of P to be $APtraces(P) = \{APtrace(\rho) \mid \rho \in runs(P)\}$.*

Definition 6. *A specification for a behavioral program P is a linear time (LT) property Φ (i.e. a subset of $(2^{AP})^\omega$). We say that P satisfies Φ, denoted $P \models \Phi$, iff $APtraces(P) \subseteq \Phi$.*

Since this definition assumes infinite runs, when dealing with systems of finite runs we pad any finite run with the trace $\emptyset^\omega$.

It is important to note, that the same set of b-threads can satisfy Φ with one event selection mechanism, and not with another. We adopt a wider perspective here, and ensure that the patched set of b-threads satisfies Φ with *all* event selection mechanisms. Such patching immediately detects and fixes any bugs that could have remained hidden with a certain mechanism, but which may emerge later. An approach that takes a specific event selection mechanism into account may also be useful for some applications.

In this work we focus on two major types of LT properties: safety properties and liveness properties. We define safety properties first, and give also the related definitions of invariants and deadlocks.

Definition 7. *An LT property Φ over AP is called a* safety *property if for all* $\sigma \in (2^{AP})^\omega - \Phi$ *there exists a finite prefix $\bar{\sigma}$ of σ such that*

$$\Phi \cap \left\{ \sigma' \in (2^{AP})^\omega \mid \bar{\sigma} \text{ is a finite prefix of } \sigma' \right\} = \phi.$$

Intuitively, a safety property states that no "bad" sequences of events may happen. Any run that causes such a sequence has a *bad prefix*; after it the run does not satisfy the property no matter how it continues.

The notion of *invariants* plays a key role in the model-checking of safety properties:

Definition 8. *An LT Φ property over AP is an* invariant *if there is a propositional logic formula φ over AP such that $\Phi = \left\{ A_0 A_1 A_2 \ldots \in (2^{AP})^\omega \mid \forall j \geq 0, A_j \models \varphi \right\}.$*

Intuitively, invariants are properties of the current state of the system, and do not reflect the history of events leading to it.

Through invariant checking one can handle *regular safety properties*: those safety properties for which the associated bad prefixes are recognizable by some finite automaton [3], or, in our case, there is a b-thread that marks its state as bad when the bad prefix is recognized. By applying the invariant model-checker to a program with these threads added, we can effectively handle general regular safety properties.

Definition 9. *We say that a (finite) run $\rho = (e_1, e_2, \ldots, e_n)$ causes a deadlock if it leads to a state s that has no enabled events (all requested events are also blocked).*

Much like invariants, deadlocks too are properties of states in the system, and not of runs.

When patching against safety violations, we will receive as input a program P and an invariant Φ. We will implicitly check that the system has no deadlocks; if it does, the patching algorithm will try to remove them. In particular, we will make sure that no new deadlocks are created while patching; otherwise we could "patch" a system by simply blocking all enabled events at its initial state.

The other type of properties we consider is liveness properties. The following is adopted from [3]:

Definition 10. *An LT property Φ over AP is called a* liveness *property if any finite word can be extended such that the resulting infinite trace satisfies Φ. Formally, let $pref(\sigma) = \{\bar{\sigma} \in (2^{AP})^* \mid \bar{\sigma}$ is a finite prefix of $\sigma\}$ and $pref(\Phi) = \bigcup_{\sigma \in \Phi} pref(\sigma)$. Then Φ is a liveness property if and only if $pref(\Phi) = (2^{AP})^*$.*

In the case of regular safety properties, invariant checking plays a key role. When it comes to liveness properties, a similar role is played by *persistence* checking:

Definition 11. *An LT property Φ over AP is called a* persistence *property if it states that a certain condition holds forever, from some point in Φ. Formally, Φ is a persistence property if there exists a propositional logic formula φ such that $\Phi = \{A_0 A_1 A_2 \ldots \in (2^{AP})^\omega \mid \exists_i$ such that $\forall_{j \geq i}, A_j \models \varphi\}$. Formula φ is called the persistence (or state) condition of Φ.*

As discussed in, e.g., [3], the model-checking of regular liveness properties is reducible to persistence checking. The latter is performed by portioning the states of the system into two sets: states in which φ holds, termed "cold" states, and states in which it does not hold, termed "hot" states. Then, the property holds if and only if there are no reachable cycles consisting strictly of hot states (which we refer to as reachable "hot cycles"). This can be checked, for instance, using a nested DFS algorithm.

When patching against liveness violations, we will receive as input a program P and a persistence property Φ. In practice, this property is given by an indicator thread that marks the system's states as either hot or cold.

5 Extending the Model-Checking of Invariants and Deadlocks

In order to prepare the ground for the correction of various safety and liveness violations, we begin by describing how to check that a behavioral program satisfies an invariant and is deadlock-free. We follow the algorithm in [3], Sect. 3.3.1, and the implementation in [9].

Any state that violates the invariant or is deadlocked is marked as "bad". We construct the state graph of the program, traverse it using DFS (trimming when arriving at a previously visited state), and check that all states reachable from the initial state are not bad. From each state we explore all enabled events (which reflects our decision to cater for all possible event selection mechanisms).

The runtime complexity of this algorithm, implemented as in [9], is as follows. Let $G = (V_G, E_G)$ denote the state graph constructed, and let n be the number of threads and $e = |E|$ the number of events in the original program. $|V_G| + |E_G|$ operations have to be performed to traverse the graph. Further, for each *state* $\in$

V_G we have to perform $n \cdot e$ operations in order to find all its enabled events. This yields:

$$T_{mc} = O\left(|E_G| + |V_G| \cdot (n \cdot e)\right).$$

This complexity is the minimum price one has to pay for running a model-checker on a behavioral program. Since our technique is based on model-checking, it will necessarily be forever linked in complexity to that of model checking [3,22], and the progress made there, for better or for worse. T_{mc} thus serves a base point with which to compare the complexity of our patching algorithms, and we are interested in how much additional overhead they incur above it.

We actually use a slightly different algorithm. For our purposes, the usual model-checking that returns a single violating run does not suffice: we want to explore *all* runs that violate the invariant or cause a deadlock.

This is achieved as follows: we traverse the state graph using the same DFS, but whenever we reach a bad state we store that information in its predecessor states. Each state already visited in the graph will thus contain information on all its bad successors. If the state is reached again, through another route from the root, we need not traverse its subtree again: we simply update the relevant states using the data already stored (see Fig. 3).

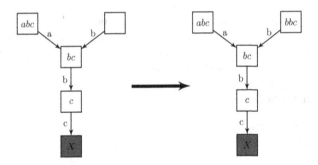

Fig. 3. When a "bad" state is reached, all its predecessors store the relative path from that point to the violation. When a node in this path is reached through a different path, the data is propagated. The DFS continues until the root stores all violating paths.

The added complexity of this algorithm is measured using the number of violating runs, Υ (OOPSilon: pun intended), and the depth of the state graph D. For each violating run we propagate at most D events to the predecessors, causing an overhead of $1 + 2 + \ldots + D$ per violating run. The total runtime complexity is thus:

$$T = T_{mc} + \Upsilon \cdot (1 + 2 + \ldots + D) = T_{mc} + O(\Upsilon \cdot D^2).$$

Finally, if all direct successors of a state are bad, then the state itself can be considered bad; this is because the patching technique we discuss will cut off the violating children, rendering the state a deadlock. We thus add the following modification: if, during the DFS, all of a state's successors are violating or deadlocked, the state itself is marked as violating; thus its successors can be ignored. The runtime worst-case complexity remains unchanged.

6   Safety Patches for Loopless Programs

6.1   Generating Linear Safety Patches

Before discussing the safety violation patching of general programs, we begin with the simpler case of finite programs that are *loopless*: their state graph contains no cycles. In a loopless program, every run is finite.

Definition 12. *A* linear safety wait-block patch *for event sequence* $(e_1, e_2, \ldots, e_n, e_{last})$, *such that* $e_{last} \in E_{prog}$, *is a b-thread with the following properties:*

- *The patch waits for events* $e_1, \ldots, e_n$, *blocks* e_{last} *once and then terminates.*
- *If the run deviates from the sequence* $e_1, \ldots, e_n$, *the patch terminates.*
- *The patch never requests events and does not label states* $(R(s) = L(s) = \emptyset$ *for all* s).

Intuitively, the patch is designed to prevent one bad run from occurring. Events $e_1, \ldots, e_n$ will be chosen according to violating runs found by the model-checker. The patch will intervene before the last event, causing another event to be triggered, thus preventing the violation.

The patch only interferes with runs starting with events $e_1, \ldots, e_n$; other runs remain unchanged. Formally:

Lemma 1 (The Locality Lemma). *Let* P *be a collection of b-threads, let* p *be a linear safety wait-block patch for event sequence* $(e_1, \ldots, e_n)$, *and let* $P' = P \cup \{p\}$ *denote the patched program. Then for any run* ρ *of* P *that does not start with events* $e_1, \ldots, e_n$, *the events of* ρ *constitute a valid run* ρ' *of* P', *and* $APtrace(\rho) = APtrace(\rho')$.

Proof. To prove that ρ' is a valid run of P', we need to show that at each synchronization point during ρ', the triggered event is also enabled; namely, it is requested and not blocked. By definition, if a run does not start with events $(e_1, \ldots, e_n)$, then the patch never requests or blocks events. Further, the original b-threads will reach the same states during ρ' as they did during ρ, consequently requesting and blocking the same events. It follows that the program P' has the same requested and blocked events as P in each state during the run. Thus, the events triggered by ρ' are enabled, and the run is indeed valid.

Finally, since the original b-threads reach the same states during ρ' as during ρ, they will have the same atomic propositions associated with them. Since the patch has no atomic propositions associated with its states, we get that $APtrace(\rho) = APtrace(\rho')$. $\square$

The Locality Lemma is our motivation for patching: it states (in this case, for linear patches) that when we add a patch to negate a single bad run, other runs remain unharmed, meaning that the patch is local. This is an advantage of our method as compared to traditional, manual, patching: our patches do not create new errors in unexpected parts of the code.

The distinct bad runs representing the bug or emanating from the new safety requirement are found by model-checking:

Linear Safety Patching(P, Φ):
1: Run the model checker on (P, Φ)
2: **if** $P \vDash \Phi$ **then**
3:    **return** P
4: $P' \leftarrow P$
5: **for** each violating run $(e_1, \ldots, e_n)$ **do**
6:    **if** $\forall i, \ e_i \in E_{env}$ **then**
7:      **return** **Failure**
8:    **else**
9:      Find the largest k such that $e_k \in E_{prog}$
10:      Create a linear safety wait-block patch p for $(e_1, \ldots, e_k)$
11:      $P' \leftarrow P' \cup \{p\}$
12: **return** P'

The idea is straightforward: the model-checker finds all runs violating Φ and we add a patch per run to prevent them. Because Φ is a safety property and P is loopless, there are only finitely many violating runs. The algorithm guarantees that the blocking performed by the patches creates no deadlocks, by first recursively marking as "bad" any state that has only "bad" children. Furthermore, because the model-checker works with respect to all possible event selection mechanisms, any bugs that emerged after the patching are fixed. The Locality Lemma guarantees that no good runs "far away" from the patch are harmed. If the algorithm returns a patched program, we thus know that it satisfies the specification Φ and causes no deadlocks.

There is also the case where the algorithm returns a failure notice, as a result of the model checker returning a violating run in which there were no program-requested events. This, of course, means that the program cannot be repaired through wait-block patching. Formally:

Lemma 2 (The Patchability Lemma). *Let P be a loopless program with state graph $G = (V_G, E_G)$ and let Φ be a safety property. Then the following three statements are equivalent:*

1. *The algorithm succeeds in returning a patched program P'.*
2. *There exist linear safety wait-block patches $p_1, \ldots, p_k$, such that $P \cup \{p_i\} \vDash \Phi$.*
3. *There exists a graph $G' = (V_G, E_{G'})$ with $E_{G'} \subseteq E_G$ and $E_G - E_{G'} \subseteq E_{prog}$, such that no states violating Φ or causing deadlocks are reachable from the initial state in G'.*

Proof. $(1) \Rightarrow (2)$ is trivial.

For (2) $\Rightarrow$ (3): Take the original state graph G, and for each p_i remove the edge corresponding to the event it blocks. Since the patched program satisfies Φ and does not deadlock, all reachable states in the graph obtained in this way satisfy Φ and do not cause deadlocks. Furthermore, by the definition of a wait-block patch, all edges removed are in E_{prog}, as needed.

For (3) $\Rightarrow$ (1): Without loss of generality, assume that P starts with an initialization event $e_{init} \in E_{prog}$. If this does not hold we can change to a new initial state s'_{init} and add a thread that forces event e_{init} to be chosen before proceeding to the original program.

Suppose that G' exists but that the algorithm returned a failure notice. We conclude that it deadlocked on the very first state, s'_{init}. This, in turn, means that state s_{init} was marked as bad, so that all paths starting in s_{init} lead to bad states. This contradicts the existence of G', thus proving the claim.   □

Condition (3) means that the original program was "not too far" from satisfying Φ: it contained some good runs and some bad runs, and through some blocking the bad runs could be averted. Observe that the equivalence of (1) and (2) is really the validity of the algorithm.

The worst case runtime complexity of the algorithm is just that of the modified model-checker, namely $T = T_{mc} + O(\Upsilon \cdot D^2)$. This shows the dependence of our algorithm on the number of violating runs in the original program. If their number and lengths are small enough our automatic patching is not much worse than regular model-checking. This also demonstrates why using this algorithm for synthesis could be costly. If the program is "far away" from satisfying Φ, as could be the case when trying to synthesize a program from scratch (say, from a general program that constantly requests all possible events), then Υ could be polynomial in the size of the state graph, greatly slowing the process.

6.2   Patching for a Specific Event Selection Mechanism

The above algorithm patches the program so that it satisfies Φ, regardless of the event selection mechanism used. However, it may be useful to patch the program for the specific mechanism M to be used, as it could speed up the patching process, reduce the number of generated patches, and most importantly, block less events, leaving open more options for further behavior refinements and repair, as explained in Fig. 4.

In this case, the model-checking algorithm is modified to return as output all violating runs of the original program, as well as all (and only) violating runs that would be created by blocking previously discovered bad transitions. Bad runs that will not be possible in the patched program, under the specific ESM, are ignored. This technique is readily applicable also to patches for programs with cycles and for liveness patches, discussed in the sequel.

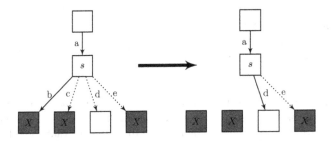

Fig. 4. In state s, a patch that considers all event selection mechanisms will block b, c, and e. A patch that considers only, say, an ESM that chooses events alphabetically, needs to block b and c, but can leave e unblocked, relying on the selection of d.

6.3   Example: Patching Tic-Tac-Toe

We demonstrate the use of the linear safety patching algorithm on the loopless Tic-Tac-Toe behavioral program from [9]. It is loopless since the fact that each step adds a new move to the board means that its state graph has no cycles.

Suppose that the original program is developed without a model-checker. At the time of development, the programmer is convinced that the program always achieves its goal, i.e., never loses (observe that this is a safety property — bad things do not happen). Various testers support this statement. The program is then deployed. Some months later, a customer defeats it and sends in the game's trace. However, the original software engineer has long quit the firm, and it would take a long time for a new engineer to repair the code. A suitable solution would be to apply an automatic patching algorithm to the malfunctioning software.

To simulate this, we took the complete program from [9], and omitted the more complex threads — those that handle situations where our opponent creates, simultaneously, two ways to win. If the human player does not try the complex strategy that create such double attacks, the program does indeed seem to work, but a skilled player can defeat it.

The automatic proof-of-concept tool is easy to use, requiring little modifications to the original program. The input is the behavioral program and the safety property Φ, given as b-threads marking bad states (e.g., victory of the opponent). The output is code files for new thread instances which are easy to read and to integrate into the original program (see Fig. 5).

Each such patch inherits from a parent class which implements its "main" function; see Fig. 6.

In our example, the patched Tic-Tac-Toe program contains 26 different patches, one of which is demonstrated in the figure. Subsequent verification by the model checker confirms that now the specification is indeed satisfied.

```
public patch1() {
    events.add(new X(2,2));
    events.add(new O(1,1));
    events.add(new X(0,0));
    events.add(new O(2,0));
}
```

Fig. 5. Example of a wait-block patch generated by the proof-of-concept tool. The patch's code contains a sequence of events that should be waited-for — events X(2,2), O(1,1) and X(0,0). The last event in the list, O(2,0), is the one that should be blocked by the patch. The automatically generated code is legible and comprehensible, as the more complicated details are hidden away in a parent class.

```
public void runBThread() {
    for (int i=0; i<events.size()-1; i++) {
        bp.bSync( none, all, none );
        if (!lastEventWas(events.get(i)))
            disablePatch();
    }
    bSync( none, all, events.getLast() );
    disablePatch();
}
```

Fig. 6. The patch thread's main function, `runBThread()` is part of the patching library, and is not added to the actual patched program. It waits for events defined by a particular patch instance (as in Fig. 5), blocking the last event and then terminating. If the events chosen deviate from those defined in the patch instance, it terminates.

7   Safety Patches for Programs with Cycles

7.1   Generating Safety Patches for Cycles

The correctness of the algorithms for linear safety patching relies on the program's state graph's having no cycles. As most reactive systems run indefinitely, periodically returning to some "idle" state, such systems cannot be patched by linear wait-block patches. For example, fixing a behavioral program that enters a bad state after a sequence of events of the form $(a)^*b$, will call for infinitely many linear patches.

Our solution is to extend the linear safety patch associated with a single sequence of events, into one that can keep track of an entire hierarchy of paths and cycles in the graph, blocking the violating event as needed.

Definition 13. *Given a state graph $G' = (V_{G'}, E_{G'})$, two special vertices marked v_{init} and v_{end} and an event $e \in E_{prog}$, a cyclic safety wait-block patch for G' is a b-thread with the following properties:*

- *It waits for all events chosen by the event selection mechanism and traverses the graph G' according to those events.*
- *Whenever state v_{end} is reached, it blocks event e once.*
- *If an event occurs such that there is no edge marked with that event, it terminates.*
- *It never requests events and does not label states.*

Intuitively, the patch is designed to prevent a family of bad runs that are similar to one another, in that they reach their bad state by transitioning from v_{end} via the event e. The graph G' will be chosen such that it contains *all* paths from v_{init} to v_{end}, thus rendering a single patch able to block that entire family of bad runs.

The Locality Lemma holds for the cyclic case as well: all runs of the original system, apart from those starting in v_{init} and ending in reaching the violating state through v_{end} and e, are valid runs of the patched system. The proof is based on the fact that in any such run, the generated patch does not request or block any events, and thus does not affect the events requested by the program.

Linear safety patches are a particular case of the cyclic ones, in which the graph G' is a path, meaning there is precisely one way to reach the violating state.

The cyclic safety patching algorithm is as follows (G denotes the full state graph traversed by the model-checker):

Cyclic Safety Patching(P, Φ):
1: Run the model checker on (P, Φ)
2: **if** $P \models \Phi$ **then**
3:     **return** P
4: **for** each violating run $(e_1, \ldots, e_n)$ **do**
5:     **if** $\forall i, \ e_i \in E_{env}$ **then**
6:         **return** **Failure**
7:     **else**
8:         Find the largest k such that $e_k \in E_{prog}$
9:         Let s_{end} denote the state reached after events $e_1, \ldots, e_{k-1}$
10:       Construct the minimal subgraph G' containing all paths in G from s_{init} to s_{end}
11:       Create a cyclic wait-block patch p for G' with states $v_{init} = s_{init}, v_{end} = s_{end}$, and event e_k.
12:       $P' \leftarrow P' \cup \{p\}$
13: **return** P'

Constructing the minimal subgraph G' is done using a modified BFS algorithm, in the following manner. Given the full graph and the two vertices s_{init} and s_{end}, we run a modified BFS search from s_{init}. Unlike a regular BFS search, where each vertex stores a single predecessor (the first vertex from which it is found), here each vertex stores *all* the vertices from which it is found. When the search is over, we begin in s_{end} and backtrack through all possible predecessors of each vertex, until reaching s_{init}. The set of edges and vertices traversed this way forms the subgraph G' that we need.

To show that every path from s_{init} to s_{end} is in G', let $p = (s_{init}, s_1, \ldots, s_n, s_{end})$ be a path. If p is simple, i.e., no state repeats itself, then clearly after $n+1$ iterations of the BFS search each vertex in p has its preceding state marked as a predecessor. Therefore, the entire path will be traversed during the backtrack phase, meaning that p is in G'.

Now, suppose that p is a complex path with one cycle (the proof for the general case is an easy extension). Then p can be expressed as follows:

$$p = (s_{init}, s_1, \ldots, s_k, \underbrace{s'_1, s'_2, \ldots, s'_j, s_k}_{\text{the cycle}}, s_{k+1}, \ldots, s_n, s_{end})$$

The states before and after the cycle are found as before. The cycle's states, $s'_1, \ldots, s'_j$, are found at the latest during the j'th iteration after the first arrival at s_k. When the cycle ends, s'_j is marked as a predecessor of s_k. Therefore, during the backtrack phase that passes through s_k, the entire cycle will be found. Consequently, the returned subgraph contains p.

To see why G' is minimal, observe that if a state is added to the subgraph it is part of at least one path from s_{init} to s_{end}, and therefore cannot be omitted from the graph.

Lemma 3. *If the algorithm returns a patched program P', then $P' \models \Phi$.*

Proof. Suppose that there exists a run ρ of P' violating Φ. Denote its states $s_1, \ldots, s_n$, and extract from them a violating run with no cycles. If $s_i = s_j$ for some $j > i$, delete states $s_{i+1}, \ldots, s_j$. Denote the remaining states as $s_{t_1}, \ldots, s_{t_k}$. The run corresponding to this state sequence was found by the model checker, and a patch for some subgraph G' which contains this run was created. Since G' contains all paths from s_1 to s_n, it also contains ρ. Therefore, the patch would have blocked the last program-requested event of ρ, causing a contradiction. $\square$

As with the linear case, it is possible for the algorithm to return a failure notice. The Patchability Lemma, which characterized programs that could be fixed in the linear case, holds for the cyclic case as well; its proof is analogous.

The complexity of the algorithm is as follows: The exploration of violating runs costs, as before, $O(T_{mc} + \Upsilon \cdot D^2)$. Constructing the relevant subgraph for each violating run costs another $|V_G| + |E_G|$ times Υ runs, yielding:

$$T = O\left(T_{mc} + \Upsilon \cdot D^2 + \Upsilon(|V_G| + |E_G|)\right).$$

Again, this shows our dependence on the number of violating runs, Υ. The smaller that number, the closer our complexity is to that of the model-checker; the higher it is, the closer we are to the notorious, worst-case complexity of the synthesis problem.

7.2 Subgraph Representation

The generated code for a linear safety patch contains only the list of events to be waited for, followed by the event to be blocked. This list can be readily understood and possibly manipulated by a human, say, for documentation or analysis. Further, the developer may simplify or generalize the patch; e.g., skip waiting for certain guaranteed events or consolidate patches into fewer "symbolic" one, using BPJ's event filters. However, when a patch traverses a complex subgraph, gaining such insights is harder. Thus, we propose to represent the subgraph as a collection of easily readable linear event scenarios, amenable to human manipulation. The operation of the cyclic safety patch will be as before.

Specifically, We use the term *line* for a finite sequence of events that occur along some contiguous path in the state graph, and along which no state is visited twice. We use the term *tail* for a line whose last event would lead to a bad state in the state graph. The program's state graph, or parts thereof, are stored as a collection of lines, each containing its sequence of events, and links to other lines that are reachable by a single event from the last event in the line. See Fig. 7.

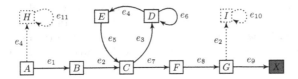

Fig. 7. A state graph of a buggy program. The model-checker returns the violating run with events e_1, e_2, e_7, e_8, e_9. The subgraph of all paths from state A to state G (see solid states and edges) is decomposed into: $line_1 = e_1, e_2$ (successors $tail, line_2$); $line_2 = e_3$ (successors $line_3, line_4$); The self-loop $line_3 = e_6$ (successors $line3, line_4$); $line_4 = e_4, e_5$ (successors $line_2, tail$); $tail = e_7, e_8$ (with event to be blocked, e_9). In addition to the run found by the model checker, the patch prevents other runs, e.g., $e_1, e_2, \underline{e_3, e_6, e_6, e_4, e_5, e_3, e_4, e_5}, e_7, e_8, e_9$, where the underlined events correspond to cycles.

Thus, each patch,

- begins by activating lines containing the initial state;
- waits for all events and traverses active lines;
- deactivates active lines when they are deviated from;
- deactivates a line and activates its successors when the line's last event occurs;
- in a tail, prior to the event leading to the bad state, blocks that event, waits for one more event, and deactivates the tail.

The line representation can be implemented in a data structure or in separate patch b-threads, each beginning with waiting for a unique activation event. This results in a number of small patches and is readily implementable in all implementations of behavioral programming.

7.3 Example: Patching a Coffee Machine

We demonstrate cyclic safety patching with a simple coffee vending machine example, which is expected to repeatedly wait for a coin, wait for a coffee request, and prepare the coffee. The main requirement is that coffee is never prepared unless a coin is first inserted. However, if immediately after power-up the user

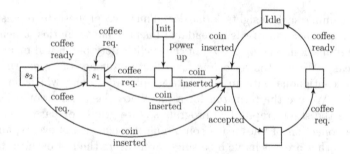

Fig. 8. The buggy coffee machine's state graph. After the `PowerUp` event, if a `CoffeeRequested` event occurs (before a coin is inserted), free coffee can be obtained infinitely many times, until a coin is inserted. The loop on the right-hand side of the graph represents the desired operation. The problematic state (marked s_1) has two enabled events: `CoffeeReady`, which is immediately requested (and selected), and the environment event `CoffeeRequested`. We expect the patch to block the `CoffeeReady` event.

requests coffee, the machine incorrectly allows coffee to be requested and prepared infinitely many times without a coin. When the first coin is inserted, the machine enters normal operation. The machine's state graph is depicted in Fig. 8.

Observe that the bug is a safety bug — coffee is served without first inserting a coin. When it is discovered and automatic patching is attempted, the first step is to have a new b-thread identify and mark bad states (namely, s_2).

The automatic patching algorithm generates a single patch, corresponding to the subgraph depicted in Fig. 9.

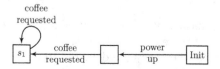

Fig. 9. The subgraph of the program's state graph for which a patch is created. It shows all paths from the graph's initial state to state s_1, in which event `CoffeeReady` must be blocked to prevent violations.

Finally, the graph of the patched program is depicted in Fig. 10, and the code generated by the proof-of-concept tool is shown in Fig. 11.

8   Dealing with Liveness

Up to this point, we dealt with safety properties — those that assert that "nothing bad happens". Another important class of properties is those involving liveness, asserting that "good things eventually happen". In this section we show how wait-block patches can be applied in order to fix liveness violations too.

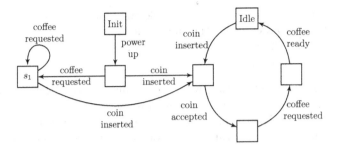

Fig. 10. The patched program's state graph (states of the patches themselves are omitted for clarity). The violating `CoffeeReady` event has been blocked, and the bad state no longer exists in the state graph.

```
public cyclicPatch1() {
    line1Events.add( new PowerUp() );
    line1Events.add( new CoffeeRequested() );
    line1 = new LineComponent( line1Events );

    line2Events.add( new CoffeeRequested() );
    line2 = new LineComponent( line2Events );

    tailEvents.add( new CoffeeReady() );
    tail = new TailComponent( tailEvents );

    line1.addSuccessor( tail );
    line1.addSuccessor( line2 );
    line2.addSuccessor( line2 );
    line2.addSuccessor( tail );

    this.addActiveComponent( line1 );
}
```

Fig. 11. The automatically-generated Java code for representation of the subgraph in Fig. 9. The first line contains events `PowerUp` and `CoffeeRequested`, and the second line contains `CoffeeRequested`. The tail contains only the event to be blocked, `CoffeeReady`. The code is readily understandable.

In the case of safety properties, ensuring that a property holds is reducible to rendering all "bad" states unreachable, and so it was straightforward to use blocking in order to correct malfunctioning programs. Recall that in Sect. 4 we mentioned that liveness violations correspond to reachable cycles of "hot" states (i.e., "hot cycles") in the program's state graph, and so it is less clear how to apply blocking. One natural approach might be to identify when the system is traversing a hot cycle, and then block one of the cycle's transitions (when an alternative exists), forcing the run to leave the cycle. This has several drawbacks:

1. Unlike in the safety case, where a bad state was never to be visited, in the liveness case it is legal to traverse the hot cycle any finite number of times. Consequently, safety-like patching would destroy good runs, which is highly undesirable.

2. Naïvely forcing the run to leave a hot cycle does not guarantee that it reaches a cold state; it could enter another hot cycle.
3. We would need to keep track of the hot cycles in the graph — the number of which could be very large.

To overcome these difficulties, we adopt a different perspective. Instead of considering runs and the hot cycles they traverse, we consider the hot states themselves. We show how, using wait-block patches, one can enforce a state-based policy that forces every run to visit cold states infinitely often, thus ensuring that the liveness property in question holds.

Our technique works by distinguishing between two types of hot states: *hot-trap* states and *hot-escapable* ones. Hot-trap states have the property that once they are visited, a liveness violation cannot be prevented; i.e., the system can never force the run into a cold state again. Consequently, hot-trap states are considered as "bad" states, and we use safety wait-block patches to render them unreachable. The hot-escapable states are those from which the system could force the run to visit a cold state, via some transitions; however, we cannot assume that these transitions may ever be traversed. In particular, it is possible for the system to continuously choose transitions that keep the run in hot states, although transitions to cold states are always enabled. We handle the hot-escapable states by enforcing *fairness*: we make sure that if a transition is enabled infinitely often, it will eventually be traversed. This type of fairness can be enforced using *probabilistic wait-block patches*, which we also call *liveness patches*. Through their use we can ensure that any liveness violations are effectively eliminated.

In the remainder of the section we discuss the liveness patching process more thoroughly.

8.1   Classifying Hot States

The first step in our repair algorithm is partitioning the hot states in the state graph into the two types mentioned. These two sets are formally defined by the algorithm below, which takes as an input the state graph of the program $G = (V_G, E_G)$, and returns the sets of hot-escapable and hot-trap states. For each hot-escapable state the algorithm also outputs its *escape-distance*, denoted δ: this is the length of a path from the hot-escapable state that reaches a cold state, and which the system can enforce regardless of the environment's behavior. See Fig. 12 for an illustration.

Classify Hot States(V_G, E_G):
1: $A \leftarrow ColdStates(V_G)$, $B \leftarrow HotStates(V_G)$, $iteration \leftarrow 1$
2: $continue \leftarrow true$
3: **while** $continue$ **do**
4:      $continue \leftarrow false$, $New \leftarrow \phi$
5:      **for** each state $s \in B$ **do**
6:          **if** at least one outgoing edge (internal or external) from s leads to a state in A, and no outgoing external edge from s leads to a state in B **then**

```
 7:        continue ← true
 8:        New ← New ∪ {s}
 9:        δ(s) ← iteration
10:      iteration + +
11:    B ← B − New, A ← A ∪ New
12:  HotEscapable ← A ∩ HotStates(V_G)
13:  HotTrap ← B
14:  return (HotEscapable, HotTrap)
```

The algorithm performs a fixpoint computation of the set A of states that are either cold or from which the system can force the execution to reach a cold state. When the algorithm terminates, this set contains the hot-escapable states.

The key point in the algorithm is line 6, which contains the condition based on which a new hot state enters A. For a state to become hot-escapable, all its external edges must lead into A, which expresses the fact that these events are beyond our control, and are controlled by the environment. Since we cannot prevent (block) them, we require that they cause no problem in the first place — namely, that they lead to states that have already been classified as hot-escapable by their being in the set A. Another condition, which handles the case where a state only has internal events enabled, is that there be at least one edge going into a state of A. The key fact is that if either condition holds, the blocking idiom can be applied to block all edges that do not lead to A.

The *escape-distance* value, δ, of a hot-escapable state indicates the number of the iteration in which it joined A. It measures the shortest guaranteeable distance to a cold state — that is, the length of the shortest such path that can be enforced by blocking.

Observe that while this algorithm serves to define hot-escapable and hot-trap states, it is not efficient — primarily because of the loop in line 5. By considering, at every iteration, only nodes that have successors that were determined hot-escapable in the previous iteration, the run time complexity can be reduced to $O(|V_G| + |E_G|)$.

8.2 Handling Hot-Trap States

As discussed previously, once the run enters a hot-trap state the system cannot guarantee that it ever reaches a cold state. Consequently, we are forced to block that entrance in the first place. This is done by applying safety wait-block patches, using the technique discussed in Sect. 7, which renders all hot-trap states unreachable.

Observe that this may remove potentially good runs too — namely, runs that go through a hot trap state yet still visit a cold state eventually. This can happen, for example, when an external event leading to another hot trap state is not triggered and, instead, an internal event that leads to a cold state is triggered. However, since we cannot depend on external events being triggered or not, our only way to ensure that no violations occur is to make hot-trap states unreachable.

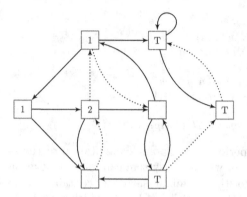

Fig. 12. Hot-trap and hot-escapable states. Hot states are marked red and contain either a number or the letter T; cold states are marked blue and are empty. Solid edges correspond to internal events, and dotted edges correspond to external events. A number inside a hot state designates the state as hot-escapable and indicates the escape-distance. The letter T designates the state as hot-trap (Color figure online).

8.3   Hot-Escapable States and Transition Fairness

The criterion used in determining the set of hot-escapable states ensures that careful use of the blocking idiom can force the run from a hot-escapable state into a cold state. The actual technique we propose is aimed at harming as few good runs as possible, and is based on *fairness*.

The notion of *fairness assumptions* [18] is used widely in formal verification, typically in order to rule out violating runs of the system because they are not realistic. Here, we discuss a special kind of fairness, called *transition fairness* [1]: if a transition is enabled infinitely often (i.e., its state of origin is visited infinitely often), then it is traversed infinitely often. We also allow a set of transitions originating from the same state to form a single constraint: if the state is visited infinitely often, then at least one of the transitions in the set is traversed infinitely often. Note that, unlike in the traditional setting where fairness is *assumed* for verification purposes, here we aim to *enforce* it within a malfunctioning system.

Intuitively, hot-escapable states have the property that if the event selection mechanism were to choose the triggered events uniformly at random, a run that visits them would eventually lead to cold states. It turns out that one can also settle for assumptions that are weaker than truly random event selection. We express these required assumptions as transition fairness constraints, and then discuss how to enforce them. Formally:

Definition 14. *Let P be a behavioral program with state graph G. A transition fairness constraint c on G is a set of one or more transitions (edges) in the graph, $\{e_1, \ldots, e_n\}$, all originating from the same node v. We say that P satisfies c, denoted $P \vDash c$, if it has the following property: if a run ρ of P visits v infinitely often, transitions from c are traversed infinitely often.*

Let $C = \{c_1, c_2, \ldots, c_k\}$ be a set of transition fairness constraints. We say that P satisfies C, denoted $P \vDash C$, if $\forall_{1 \leq i \leq k} P \vDash c_i$.

We now define a set of specific transition constraints for each of the hot-escapable states in the graph, and then show that they suffice for guaranteeing the liveness property in question.

Definition 15. *Let P be a behavioral program with state graph $G = (V_G, E_G)$, and let $V_{hot\text{-}escapable} \subseteq V_G$ be its set of hot-escapable states with respect to some liveness property Φ. For each $v \in V_{hot\text{-}escapable}$, the transition fairness constraint of v, $\tau(v)$, is defined as follows:*

- *if v has transitions corresponding to external events, $\tau(v)$ is the set of these transitions.*
- *otherwise, v has a neighbor, u, such that u is a cold state or $\delta(u) < \delta(v)$. In this case, we define $\tau(v)$ to be the edge leading from v to u.*

Observe that for every hot-escapable state v, $\tau(v)$ can be found during the hot state classification algorithm at no additional cost. We define the set of transition fairness constraints of the entire program to be the set of transitions fairness constraints on all its hot-escapable states, namely $\tau(P) = \bigcup_{v \in V_{hot\text{-}escapable}} \tau(v)$. The following proposition justifies our choice of constraints:

Lemma 4. *Let P be a behavioral program and let Φ be a liveness property. If P has no hot-trap states with respect to Φ and $P \vDash \tau(P)$, then $P \vDash \Phi$.*

Proof. Suppose, towards contradiction, that $P \nvDash \Phi$. Then there exists a run ρ of P and a hot state $v_0 \in V_{hot}$, such that v_0 appears infinitely often in ρ. Since P has no hot-trap states, v_0 is hot-escapable.

By our assumption that the constraints of $\tau(P)$ hold, there exists a neighbor of v_0, denoted v_1, that also appears infinitely often in ρ, and this v_1 is either a cold state or a hot-escapable state with $\delta(v_1) < \delta(v_0)$. If the former holds, then $\rho \vDash \Phi$ and we are done. If the latter holds, we reapply the same logic iteratively. Clearly, this produces a chain of hot-escapable states $v_0, v_1, \ldots, v_n$, all appearing infinitely often in ρ, with $\delta(v_0) > \delta(v_1) > \ldots > \delta(v_n)$. Since $\delta(v_0)$ is finite, this process ends in visiting a cold state infinitely often, again implying that $\rho \vDash \Phi$. □

Note that the lemma assumes that P has no hot-trap states. However, this is not a real limitation, since, as previously explained, we can first apply safety patching to make such states unreachable.

8.4   Liveness Patches

We have characterized fairness constraints that are sufficient for correcting the liveness violation. Unfortunately, behavioral programs are not guaranteed to be fair. This is an intrinsic property of the event selection mechanisms commonly used in behavioral programming. For example, in arbitrary or priority-based selection certain transitions might be enabled infinitely often but never triggered. Consequently, we introduce a new type of patch, termed a *liveness wait-block patch*, aimed at enforcing a transition fairness constraint on the program.

Definition 16. *Given a state graph* $G = (V_G, E_G)$, *a probability* η, *a hot-escapable state* $v \in V$ *and its transition fairness constraint* $\tau(v)$, *a liveness wait-block patch for* v *is a b-thread with the following properties:*

- *It waits for all events chosen by the event selection mechanism and traverses the graph* G *according to them.*
- *It keeps track of the present state and notes when the execution reaches* v.
- *Whenever in* v, *with probability* $1 - \eta$ *the patch does nothing. With probability* η, *it blocks all transitions except those in* $\tau(v)$.
- *It does not request events and does not label states.*

Intuitively, liveness wait-block patches are a way of incrementally injecting fairness into specific states of an already existing program, without modifying existing code. When the patch is applied to a hot-escapable state, it enforces the fairness constraint of that state; in runs in which the state is visited infinitely often, at least one of the transitions specified by the constraint will be triggered infinitely often. Indeed, the probability that edges in $\tau(v)$ are not traversed after m visits to v approaches 0 as m tends to infinity, and this is true even for very small values of η. Note that, despite their probabilistic nature, these are essentially wait-block patches: they wait for a sequence of events, and then apply blocking to steer the run in the right direction.

Observe that it is indeed always possible to block all the transitions except those in $\tau(v)$. The only events that cannot be blocked are the external ones; and if there are external transitions in v, they are all in $\tau(v)$ by definition. Further, observe that by their definition liveness patches cannot cause deadlocks in states that were deadlock-free before the patching — as the patch always leaves unblocked at least one event that was already enabled.

Our motivation for using probability-based blocking is the desire to leave good runs unaffected. Choosing η to be small still guarantees that the fairness constraint holds, but makes it likely that runs that scarcely visit the state remain unaffected.

As in the case of cyclic safety patches (Sect. 7.2), liveness patches can be represented as a collection of *lines* and *tails* to make them more comprehensible.

We point out that so far we have dealt strictly with deterministic behavioral programs. Our probabilistic liveness patches, however, introduce nondeterminism into the system. This nondeterminism is not "against the grain" of behavioral programming, and indeed, extending behavioral programming definitions to support nondeterminism is straightforward, and is omitted.

8.5   The Liveness Patching Algorithm

Based on the discussion in the previous sections, we now present the patching algorithm itself:

Liveness Patching(P, Φ):

1: $P' \leftarrow P$
2: Run the model checker on (P, Φ)
3: **if** $P \models \Phi$ **then**
4:     **return** P
5: Run algorithm *Classify Hot States* on the state graph
6: **for** each hot state $v_h \in V$ **do**
7:     **if** v_h is a hot-trap **then**
8:         **if** creating a safety patch to prevent runs from reaching v_h is impossible **then**
9:             **return** **Failure**
10:        Create a safety patch $p^s_{v_h}$ that prevents runs from reaching v_h
11:        $P' \leftarrow P' \cup \{p^s_{v_h}\}$
12:    **else**
13:        Create a liveness patch $p^\ell_{v_h}$ for v_h
14:        $P' \leftarrow P' \cup \{p^\ell_{v_h}\}$
15: **return** P'

Observe that, as in the safety patching algorithm of Sect. 7, it may be impossible to create safety patches for hot-trap states in certain cases. One extreme example is when the entire state graph consists of hot-trap states only, so that attempting to render these states unreachable produces a trivial program that deadlocks in its initial state. In such cases, the algorithm returns a failure notice.

The correctness of the algorithm is established by the following lemma:

Lemma 5. *Let P' be a patched program returned by the algorithm, and let ρ be a run of P'. Then with probability 1, $\rho \models \Phi$.*

Proof. By the previously proved correctness of safety patching (Lemma 3), the algorithm ensures that there are no reachable hot-trap states in P'. By Lemma 4, it suffices to show that P' satisfies the constraints in $\tau(P)$ with probability 1. This claim is immediately derived from the definition of a liveness wait-block patch (Definition 16) and the discussion following it.   □

Part of our motivation for using wait-block patches in repairing violated safety properties was the Locality Lemma, which stated that any good runs remain unchanged. Unfortunately, that lemma cannot be proved for the liveness case; in fact, any liveness wait-block patch, by definition, might affect good runs as well as bad ones. We settle for the following:

Lemma 6 (The Weak Locality Lemma (Liveness)). *Let P be a collection of b-threads, let p be a liveness wait-block patch for hot-escapable state s_h, and let P' denote the patched program $P \cup \{p\}$. Any run ρ of P that does not reach s_h constitutes a valid run ρ' of P', and $APtrace(\rho) = APtrace(\rho')$.*

The proof is similar to that of the safety case and is omitted. The result is weaker, in the sense that if s_h is hot-escapable then good runs that pass through it might, with some probability, become invalid in the patched program. That probability increases the more times they pass through s_h. However, the effect on good runs can be reduced by decreasing the patches' blocking probability η.

The complexity of the liveness patching algorithm is as follows. The model checking phase costs $O(T_{mc})$. Classifying the hot states is linear in the size of the state graph. Each hot-trap state is then handled as a safety violation, adding $O(|V_{hot\text{-}trap}| \cdot (D^2 + |V_G| + |E_G|))$. Finally, for every hot-escapable state, we must construct the sub-graph needed to check when it is visited, yielding another $O(|V_{hot\text{-}escapable}| \cdot (|V_G| + |E_G|))$. Combining these, the total worst-case complexity becomes:

$$T = O\left(T_{mc} + (|V_{hot}| + 1) \cdot (|V_G| + |E_G|) + |V_{hot\text{-}trap}| \cdot D^2\right).$$

The runtime complexity shows the algorithm's dependence on the number of hot states: the smaller it is, the closer our complexity is to that of regular model-checking. In the next section we discuss a heuristic-based approach for reducing this in practice.

8.6   Minimal Fairness Enforcement

In the algorithm just discussed, we first rendered all the hot-trap states in the state graph unreachable and then enforced a set of transition fairness constraints in the hot-escapable states. By Lemma 5, we are guaranteed that this repairs any liveness violations in the program. However, the number of enforced fairness constraints is rather large — it is approximately the number of hot-escapable states in the program. Despite this, in some cases one can settle for fairness constraints that are far less extensive. For instance, consider the graph in Fig. 13.

Decreasing the number of fairness constraints being enforced is highly desirable, for two reasons. First, as we mentioned earlier, we wish to perform as few modifications to the original program as possible, and enforcing fewer constraints clearly serves this goal. Second, the size of the automatically generated

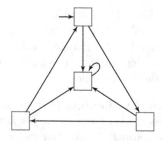

Fig. 13. An instance where the liveness repair algorithm would enforce more fairness than is required. The graph above has three hot-escapable states, and the algorithm would enforce transition fairness on the three edges leading from them into the cold state. Clearly, it suffices to settle for just one of these three constraints in order to guarantee that the cold state is eventually reached. Similar, larger constructions show that our algorithm might enforce any number of fairness constraints where just one would suffice.

code module is in correlation with the number of constraints that it enforces. Hence, fewer constraints means shorter modules, which are easier to maintain.

A natural question thus arises: can one identify a minimal-size set of fairness constraints that need be enforced on a given behavioral program in order to ensure that a given liveness property holds? Formally, we define the minimal fairness problem MF_{opt}, as follows: Given a behavioral program P with state graph $G = (V_G, E_G)$ and a liveness property Φ, such that G has no hot-trap states with respect to Φ, find a minimal-size set C of transition fairness constraints such that $P \vDash C \Rightarrow P \vDash \Phi$.

Unfortunately, it turns out that this problem is NP-complete. In [5], the authors study the problem of synthesis in the face of incomplete knowledge about the system's environment. In particular, they show that the problem of finding a minimal fairness assumption on the environment in order to make a given specification realizable is NP-complete. It is straightforward to show that this problem is reducible to MF_{opt} and that MF_{opt} is in NP, rendering it NP-complete.

Given this fact, we propose a greedy algorithm for approximating MF_{opt} in practice. The algorithm starts with an empty constraint set, and adds new constraints iteratively, in a manner similar to the way algorithm *Classify Hot States* finds the set of hot-escapable states.

Throughout its iterations, the algorithm maintains a growing set of already "handled" hot-escapable states. This is the set of states for which enforcing the current set of fairness constraints guarantees that they partake in no liveness violations. In other words, a run that visits any of these states infinitely often will reach a cold state infinitely often too.

The set of handled states is increased in each iteration. There are two ways for a state v to become handled:

1. By direct fairness enforcement: this happens when the algorithm chooses to enforce a fairness constraint leading from v into the set of already handled states.
2. By indirect domination: if, due to previous fairness constraints, all of v's successors are already handled, then v itself can be immediately marked as handled.

At each iteration, the algorithm imposes one fairness constraint, meaning that precisely one vertex becomes handled through method 1. Our criteria in choosing this particular vertex is trying to maximize the number of states that will become handled through method 2. The actual choice is performed by looking at all the candidates, namely nodes that can become handled through the enforcement of a single constraint. Each candidate is then assigned a value, which is its number of hot-escapable predecessors that are not yet handled (observe that these predecessors are precisely the vertices with potential to become dominated by choosing this vertex). Finally, the highest valued candidate is selected, and the corresponding fairness constraint is enforced. Here is pseudo-code outline of the algorithm:

Approximate $MF_{opt}(V, E)$:
1: $A \leftarrow HotStates(V)$, $handled \leftarrow \phi$, $constraints \leftarrow \phi$
2: **while** $A \neq \phi$ **do**
3:     $candidates \leftarrow FindAllCandidates()$
4:     $max \leftarrow MaxValuedCandidate()$
5:     Add a constraint that handles max to $constraints$
6:     Move max to from A to $handled$
7:     **while** there are nodes in A dominated by $handled$ **do**
8:         Move dominated nodes from A to $handled$
9: **return** $constraints$

The two subroutines, $FindAllCandidates$ and $MaxValuedCandidate$, are omitted. As with algorithm *Classify Hot States*, an efficient implementation of the algorithm and its subroutines runs in time that is linear in the size of the program's state graph.

8.7   Example: Liveness Patching for the Dining Philosophers

We implemented our liveness patching algorithm (including the greedy approximation algorithm) within our proof-of-concept tool. For evaluation, we used the dining philosophers problem [7]. A behavioral implementation thereof includes the events of a philosopher picking up and putting down a given fork, a b-thread for the behavior of each philosopher and a b-thread for each fork. Each philosopher's b-thread is subject to a strict event sequence: pick up one fork, pick up the other, put down one fork, put down the other. Each fork's b-thread waits for events that change its state, and blocks illegal events (e.g., a second picking up, or, a putting down by the "wrong" philosopher). In [9] we model-checked this problem and variations thereof for safety and liveness properties.

For our experiment, we used a variant where the first $n - 1$ philosophers are left-handed and the last one is right handed, which prevents deadlocks. All events in the program are internal, and so no hot-trap states exist. Finally, the liveness property used was this: "Philosopher #1 eats infinitely often". The results are shown in Table 1.

Each fairness constraint is translated into the actual code that enforces it, using the same mechanism as for safety patches. Although there may be many patches (as the example demonstrates), each of them is fairly comprehensible. The possibly high number of patches was part of our motivation for using the greedy algorithm; coming up with better algorithms to further reduce this number remains a topic for future work.

9   Limited-Depth Repair

9.1   Automatic Repair from Field Error Reports

Many facilities exist for end-users to send reports of software failures to the software vendor (see, e.g., Fig. 14). Typically, these reports correspond to violated safety properties (e.g., "the system never crashes").

Table 1. Comparing the results of the naïve repair algorithm and the greedy approximation repair algorithm for the dining philosophers problem, with 9–12 philosophers. The *States* column shows the total number of states in the program. The *Patches (Naïve)* and *Patches (Greedy)* columns show the number of patches generated by the naïve and greedy algorithms, respectively. Observe that since the naïve algorithm generates one fairness constraint per hot-escapable state, the *Patches (Naïve)* column reflects the number of hot-escapable states as well. Finally, the *Reduction* column shows the percentage of patches saved by using the greedy version.

#Philosophers	#States	#Patches (Naïve)	#Patches (Greedy)	Reduction
9 Philosophers	19682	17495	9913	43 %
10 Philosophers	59048	52487	30760	41 %
11 Philosophers	177146	157463	93989	40 %
12 Philosophers	531440	472391	287283	39 %

Fig. 14. Event logs from bug reports are used in patch construction.

For behavioral programs, we propose a methodology for using such failure reports in order to cope with the state-explosion problem inherent to model-checking, and to patch programs with many violating runs:

– The failure report contains an event log.
– Using the fact that the effect of a patch is local, we constrain the model checking depth to a neighborhood of the path of the failure (the bad run), followed by a limited fan-out of possible continuations, past the blocked transition.
– This is enforced by a dedicated b-thread, which monitors all events, and when an event occurs that is not along the reported bad path, it starts counting the distance from the bug report. When the distance is greater than a given parameter, the b-thread calls a model-checker API to prune the search.
– Finally, the safety patch is generated as above.

Such patching prevents the failure reported by the end-user, along with any other failures "not far" from it, and can help when full model-checking and patching consumes too much resources. The search-depth parameter is key, and needs to be adjusted per repaired program; higher depth means repairing more violations, but poorer performances. It is up to the user to use knowledge of the program's state graph, or run tests, in order to come up with the best choice.

9.2   Example: Limited-Depth Repair of the Dining Philosophers

Again consider the dining philosophers problem [7], this time where all philosophers are left handed. The reported bug fixed is the classical deadlock where all philosophers pick up the fork on their left. Table 2 shows the results of patching for the single bad run that we gave the patcher.

Table 2. Patching the dining philosophers problem using bounded depth patching. Receiving a bug report (e.g., each philosopher picked up a single fork), the algorithm searches for event sequences that deviate from, or continue, the event trace in the bug report by no more events than the search depth parameter. The patches handle cycles discovered within the search depth (e.g., one of the philosophers completing a full cycle of picking up and putting down her two forks, while the others do not proceed). The tests were carried out on a PC with a Intel Quad Core Q6600 CPU @ 2.40 GHz.

Search depth	3 Philosophers	6 Philosophers	9 Philosophers
3	3 patches	1 patch	1 patch
	3 loops	2 loops	2 loops
	0.5 s	4.2 s	30 s
4	15 patches	2 patches	3 patches
	30 loops	4 loops	6 loops
	1.2 s	22 s	4.5 min
5	20 patches	12 patches	12 patches
	380 loops	1200 loops	2580 loops
	3.2 s	2 min	45 min

10   Related Work

The research in [17, 20, 21] presents fault localization and automatic repair of programs, where a set of software components that are suspected to cause a fault is replaced by a set of synthesized components, such that the resulting system is guaranteed to meet the full specification. Automatic repair of concurrency bugs (e.g., accessed to shared memory), is presented in [16]. The detection mechanism uses bad runs associated with bug reports, and the analysis involves actual execution. The repair is manifested in modification to existing code. Genetic-programming-based repair of legacy C programs is demonstrated in [24]. The repair relies on changes to existing code in order to correct problems that were assumed to be local in nature. In [2], genetic-programming is combined with co-evolution of the test cases against which the program is evaluated. Naturally, any work on automatic-repair would be considered a particular case of program synthesis [4, 19].

As mentioned in Sect. 8.6, our work on repairing liveness violations relates to that of [5], where the authors show how to synthesize fairness assumptions the environment must uphold in order for a specification to be realizable. Our work tackles similar difficulties, but in a different setting: the environment is fixed, and fairness constraints on the system are synthesized. Our repair algorithm then imposes those constraints on the previously designed system.

As for other approaches for coordinating simultaneous behaviors, such as Esterel, BIP or Linda (see related work in [11,12] for a comparison of behavioral programming with these approaches), we believe that comparable localized repair mechanisms would be possible. The key would be implementing the equivalent of blocking which, combined with ability to subscribe to all events, is central to our solution. This, of course, is possible, as it was in Java and Erlang, and could also benefit other aspects of incremental development in these environments.

11   Conclusion and Next Steps

The contribution of the present paper is in the proposed automated approach, in which faulty components are neither identified nor modified. Instead, the system is non-intrusively augmented with additional components, to yield desired overall system behaviors. The entire approach is made possible by the incrementality and modularity of behavioral programs. The new components are readily understandable by humans, and can be documented, enhanced, or generalized as part of standard development. The generated patches can then be distributed to users without re-distributing the original software. Finally, contributing to the on-going and up-hill battle with state explosion, we propose a methodology and a practical technique for constructing local patches using limited-depth model-checking.

This research is a step in the direction of developing methodologies and tools for the repair of behavioral programs. An important next step is to enrich the tool with interactive capabilities, allowing the developer to examine the state graph and enhance the proposed repairs: consolidating similar patches, generalizing or constraining patch functionality, or perhaps changing existing code after all.

Future research problems include repairing the program with regard to time-related properties, as well as integration with other formal methods tools and techniques, including other synthesis algorithms, symbolic model-checking, and compositional verification. Our tool could be combined with Java Pathfinder [23] or other tools to explore support of richer inter-process communication beyond solely behavioral events, and possibly solving concurrency problems among b-threads, as in [16].

We hope that with further developments in incremental, non-intrusive development, supported by powerful repair automation, the task of software maintenance may eventually shed its present (often lackluster) image, becoming a rewarding undertaking, allowing software engineers to quickly address customer needs in a productive, satisfying manner.

Acknowledgments. We thank A. Kantor, S. Maoz, Y. Sa'ar, S. Szekely and G. Wiener for their valuable suggestions on the manuscript. The research of D. Harel, G. Katz and A. Marron was supported by The John von Neumann Minerva Center for the Development of Reactive Systems at the Weizmann Institute of Science, by an Advanced Research Grant from the European Research Council (ERC) under the European Community's 7th Framework Programme (FP7/2007–2013), and by the Israel Science Foundation. The research of G. Weiss was supported by the Lynn and William Frankel Center for CS at Ben-Gurion University, by a reintegration (IRG) grant under the European Community's FP7 Programme, and by the Israel Science Foundation.

References

1. Aminof, B., Ball, T., Kupferman, O.: Reasoning about systems with transition fairness. In: Baader, F., Voronkov, A. (eds.) LPAR 2004. LNCS (LNAI), vol. 3452, pp. 194–208. Springer, Heidelberg (2005)
2. Arcuri, A., Yao, X.: A novel co-evolutionary approach to automatic software bug fixing. In: Proceedings of the 10th IEEE Congress on Evolutionary Computation (CEC), pp. 162–168 (2008)
3. Baier, C., Katoen, J.-P.: Principles of Model Checking. MIT Press, Cambridge (2008)
4. Bloem, R., Jobstmann, B., Piterman, N., Pnueli, A., Sa'ar, Y.: Synthesis of reactive(1) designs. J. Comput. Syst. Sci. **78**, 911–938 (2012)
5. Chatterjee, K., Henzinger, T.A., Jobstmann, B.: Environment assumptions for synthesis. In: van Breugel, F., Chechik, M. (eds.) CONCUR 2008. LNCS, vol. 5201, pp. 147–161. Springer, Heidelberg (2008)
6. Damm, W., Harel, D.: LSCs: breathing life into message sequence charts. J. Form. Methods Syst. Des. **19**(1), 45–80 (2001)
7. Dijkstra, E.W.: Hierarchical ordering of sequential processes. Acta Inf. **1**, 115–138 (1971)
8. Harel, D., Kugler, H., Marelly, R., Pnueli, A.: Smart play-out of behavioral requirements. In: Aagaard, M.D., O'Leary, J.W. (eds.) FMCAD 2002. LNCS, vol. 2517, pp. 378–398. Springer, Heidelberg (2002)
9. Harel, D., Lampert, R., Marron, A., Weiss, G.: Model-checking behavioral programs. In: Proceedings of the 11th International Conference on Embedded Software (EMSOFT), pp. 279–288 (2011)
10. Harel, D., Marelly, R.: Come, Let's Play: Scenario-Based Programming Using LSCs and the Play-Engine. Springer, Berlin (2003)
11. Harel, D., Marron, A., Weiss, G.: Programming coordinated behavior in Java. In: D'Hondt, T. (ed.) ECOOP 2010. LNCS, vol. 6183, pp. 250–274. Springer, Heidelberg (2010)
12. Harel, D., Marron, A., Weiss, G.: Behavioral programming. Commun. ACM **55**(7), 90–100 (2012)
13. Harel, D., Marron, A., Weiss, G., Wiener, G.: Behavioral programming, decentralized control, and multiple time scales. In: Proceedings of the SPLASH Workshop on Programming Systems, Languages, and Applications Based on Agents, Actors, and Decentralized Control (AGERE!), pp. 171–182 (2011)
14. Harel, D., Pnueli, A.: On the development of reactive systems. In: Apt, K.R. (ed.) Logics and Models of Concurrent Systems. NATO ASI Series, vol. F-13. Springer, New York (1985)

15. Harel, D., Segall, I.: Planned and traversable play-out: a flexible method for executing scenario-based programs. In: Grumberg, O., Huth, M. (eds.) TACAS 2007. LNCS, vol. 4424, pp. 485–499. Springer, Heidelberg (2007)

16. Jin, G., Song, L., Zhang, W., Lu, S., Liblit, B.: Automated atomicity-violation fixing. In: Proceedings of the ACM SIGPLAN Conference on Programming Language Design and Implementation (PLDI) (2011)

17. Jobstmann, B., Griesmayer, A., Bloem, R.: Program repair as a game. In: Etessami, K., Rajamani, S.K. (eds.) CAV 2005. LNCS, vol. 3576, pp. 226–238. Springer, Heidelberg (2005)

18. Manna, Z., Pnueli, A.: The Temporal Logic of Reactive Systems: Specification. Springer, New York (1992)

19. Pnueli, A., Rosner, R.: On the synthesis of a reactive module. In: Proceedings of the 16th ACM Symposium on Principles of Programming Languages (POPL), pp. 179–190 (1989)

20. Staber, S., Jobstmann, B., Bloem, R.: Diagnosis is repair. In: Proceedings of the 16th International Workshop on Principles of Diagnosis, pp. 169–174 (2005)

21. Staber, S., Jobstmann, B., Bloem, R.: Finding and fixing faults. In: Borrione, D., Paul, W. (eds.) CHARME 2005. LNCS, vol. 3725, pp. 35–49. Springer, Heidelberg (2005)

22. Valmari, A.: The state explosion problem. In: Reisig, W., Rozenberg, G. (eds.) Lectures on Petri Nets I: Basic Models: Advances in Petri Nets. LNCS, vol. 1491, pp. 429–528. Springer, Heidelberg (1998)

23. Visser, W., Havelund, K., Brat, G., Park, S., Lerda, F.: Model checking programs. Autom. Softw. Eng. 10, 203–232 (2003)

24. Weimer, W., Forrest, S., Le Goues, C., Nguyen, T.: Automatic program repair with evolutionary computation. Commun. ACM 53, 109–116 (2010)

25. Wiener, G., Weiss, G., Marron, A.: Coordinating and visualizing independent behaviors in Erlang. In: Proceedings of the 9th ACM SIGPLAN Erlang Workshop (2010)

Designing Adaptive Systems Using Teleo-Reactive Agents

Graeme Smith[1]([✉]), J.W. Sanders[2,3], and Kirsten Winter[1]

[1] School of Information Technology and Electrical Engineering,
The University of Queensland, Brisbane, Australia
smith@itee.uq.edu.au
[2] African Institute for Mathematical Sciences (AIMS), Cape Town, South Africa
[3] Department of Mathematical Sciences, Stellenbosch University,
Stellenbosch, South Africa

Abstract. Although adaptivity is a central feature of agents and multi-agent systems (MAS), there is no precise definition of it in the literature. What does it mean for an agent or for a MAS to be adaptive? How can we reason about and measure the ability of agents and MAS to adapt? How can we systematically design adaptive systems? In this paper, we provide a formal definition of adaptivity, and a framework for designing adaptive systems aimed at addressing these issues.

The definition of adaptivity, based on Dijkstra's notion of self stabilisation, is independent of any particular mechanism for ensuring adaptivity, and any particular specification notation. The framework for designing adaptive systems is similarly independent of both implementation mechanisms and specification notation. It is based on the paradigm of teleo-reactive agents proposed by Nilsson: a paradigm in which agents move towards their goal in the presence of a continually changing environment.

1 Introduction

Adaptivity is central to the functionality of many multi-agents systems (MAS) [36]. Agents adapt gradually to their environment using, for example, machine learning techniques [20], and the distributed nature of MAS is exploited to make them robust against both external disturbances and agent failures. The inherent complexity of such systems has led to much interest in systematic approaches to their development in the area of agent-oriented software engineering (AOSE) [37], and growing interest in the use of formal methods to complement assurance approaches based primarily on simulation [6,17].

These development and assurance approaches rely on precise behavioural specifications of the agents or MAS being engineered. Such specifications can be developed in an *ad-hoc*, system-by-system fashion, but a more systematic approach is hindered by the lack of a precise definition of adaptivity and a general framework for specifying adaptive behaviour. As Mohyeldin *et al.* [21] state

R. Kowalczyk and N.T. Nguyen (Eds.): TCCI XVI, LNCS 8780, pp. 34–61, 2014.
DOI: 10.1007/978-3-662-44871-7_2

"Still open is a semantically well defined process to design adaptive systems, while concentrating on the adaptive behaviour rather than discussing implementation details."

In this paper, we address this issue by

1. providing a formal definition of adaptivity which is independent of any particular adaptivity mechanism and general enough to use with any specification notation, and
2. based on this definition, a framework for specifying the behavioural requirements of adaptive agents and MAS in a systematic way.

We begin in Sect. 2 by motivating our definition of adaptivity (which is based on our previous work [26,29]) with respect to representative informal notions of adaptivity arising in the literature. We then provide a formal model of MAS in Sect. 3 as a basis for formalising our adaptivity definition in Sect. 4. In Sect. 5, we introduce a specification framework for modelling adaptive systems based on Nilsson's teleo-reactive agents paradigm [23,24]. The approach is applied to a case study in Sect. 6 before we conclude in Sect. 7.

2   What is Adaptivity?

Various informal notions of adaptivity can be found in the literature. Below are some representative examples.

- *Adaptivity should allow change in system functionality* [13]. The view is that, by adapting to environmental change, a system or agent may offer new operations on new states.
- *Adaptivity should include self-organisation* [12,16]. Systems should be able to reconfigure to adjust for changes in the environment or agent capabilities.
- *Adaptivity should include self-optimisation* [3]. In response to a change in externally-set parameters (*i.e.*, global variables) the agents are able spontaneously and autonomously to perform calculations (viewed as optimising certain local variables) which result in the system returning to a desired state.
- *Adaptivity should include (machine) learning* [20,33]. It should, for example, enable improved response to change so that when confronted with the same situation in the future, the system or agent adapts more quickly and efficiently.

Our goal is to find a general definition which covers each of the notions above.

The most fundamental feature of an adaptive system is its ability to change behaviour (*i.e.*, functionality) in response to environmental change. The prospect of changing behaviour at first conjures up visions of systems which are somehow more advanced than standard computer programs. However, changing behaviour is illusory since a system, when viewed at a certain level of abstraction as a simple state machine, does not actually change what it is capable of doing. As shown in [13], it simply moves to a new state in which different actions and environmental interactions are possible. This is true even of approaches to evolutionary computing [10] and machine learning [20]: underlying such systems is just a computer program. Given this observation, what distinguishes an adaptive system from a merely reactive one? Indeed, a number of approaches for adaptive systems seem to support designs which respond to environmental change without further consideration for what makes those designs adaptive [4,5].

To answer this question, we appeal to the notion of *legitimate states* of a system introduced by Dijkstra [8]. In this work, 'legitimacy' is defined by a state invariant capturing those states in which the system behaves as intended. A typical reactive system would operate only in such legitimate states changing its behaviour to reflect environmental interactions and, importantly, always operating as intended.

Dijkstra's paper is concerned however with self-stabilisation of distributed systems. His examples consist of token ring networks in which a legitimate state is one in which there is exactly one token present. He presents several algorithms which, given an arbitrary number of tokens initially, end up in a legitimate state.

An important feature of Dijkstra's self-stabilising networks is that even if they start in states which are not legitimate states, they are guaranteed to reach legitimate states after a finite number of system actions. A typical reactive system may not operate at all when not in a legitimate state. Also, it is not designed to perform actions which allow it to reach a legitimate state. An adaptive system, on the other hand, is required to be a reactive system which, like Dijkstra's self-stabilising token rings, can reach legitimate states from illegitimate ones.

In other words, an adaptive system is one which, when placed in a particular environment, has a defined set of legitimate states and when in an illegitimate state reaches a legitimate state again. Indeed, the period before the system reaches the legitimate state is the time when the system is adapting.

Following Dijkstra's definition, we do not allow a system in a legitimate state to enter an illegitimate state of its own accord. That is, defined transitions of the system from legitimate states enter only other legitimate states. A system is placed in an illegitimate state by an *external* action, *i.e.*, an action that is not regarded as part of the system's specification. This action may represent a change in the environment due to an unforeseen disturbance, or the passing of a threshold point in an environment that is gradually changing over time. Alternatively, an external action may represent a change to the system itself. In the case where the system is an agent, this may be caused, for example, by the action of a software virus changing internal data. In the case where the system is a MAS, it may be caused by the failure of a component agent.

In each case we can reason about adaptivity as the ability to 'recover' from the external event, *i.e.*, the ability of the system to reach a legitimate state. Since systems will, in general, be adaptive to only a subset of all possible external actions, we qualify our definition of adaptivity with respect to a specific external action. Furthermore, we quantify adaptivity with respect to the number of actions required for the system to adapt. At a given level of abstraction, this provides us with a metric for comparing different adaptivity mechanisms.

Dijkstra's systems are *closed* in the sense that they do not interact with an external environment. The notion of legitimate states must for our purposes be extended to *open* systems by considering the state to include both that of the system and its environment. Then our definition of adaptivity covers each of the informal definitions above.

- *Adaptivity should allow change in system functionality.* Since the environment state is part of each legitimate state, a change in the environment will result in a change in the system states which constitute a legitimate state. Moving to such a system state will result in a different behaviour (and hence functionality).
- *Adaptivity should include self-organisation.* Self-organisation of a system can be seen as moving towards a legitimate state. Indeed, the approaches of both Güdemann *et al.* [16] and Georgiadis *et al.* [12] are based on satisfying certain system constraints, or invariants. Using the terminology of complex systems, self-organisation may be viewed as autonomous convergence to attracting states [25]. Under our definition the attracting states are the legitimate states.
- *Adaptivity should include self-optimisation.* With respect to self-optimisation, our definition would identify optimal states with legitimate states. While this shows the applicability of our definition in theory, it may not always be possible to specify the optimal states in practice.
- *Adaptivity should include (machine) learning.* In supervised machine learning (positive and negative) examples of a concept Q are provided to enable subsequent approximate classification of specimens into those satisfying Q and those not satisfying Q. The set of legitimate states expresses approximate classification of Q (for example as formalised by Valiant in probably approximately correct (PAC) learnability [35]). Initialisation of the learning protocol is regarded as an external action, and the agent 'learns Q' if, and only if, it reaches a legitimate state after such initialisation.
  Our quantification of adaptivity with respect to the number of actions required for the system to adapt enables us to specify improved response to external actions.

3   A Formal Model of Multi-Agent Systems

In order to formalise our definition of adaptivity, we begin by providing formal representations of agents and MAS. Since we consider agents as being artifacts that are realised by software, they can – on a low level of abstraction – be represented as labelled transition systems (LTS). An LTS comprises a (possibly infinite) set of states, a (possibly infinite) set of initial states, and a collection of actions which cause (possibly nondeterministic) state transitions. Similar concepts have been used in the agent literature before. For example, formalisms with an underlying transition systems semantics such as Z and Object-Z have been suggested for modelling agents and MAS [9,14]. Also, Hunter and Delgrande [18] use transition systems which they extend with a metric function to capture "plausibility" amongst belief states. Without loss of generality, we assume such a metric can be encoded in the transition system.

Definition 1. *An LTS is a 4-tuple* $S = (Q, I, \Sigma, \delta)$ *where*

- *Q is the (possibly infinite) set of states.*
- *$I \subseteq Q$ is the non-empty set of initial states.*
- *Σ is the set of actions (or labels).*
- *$\delta \subseteq Q \times \Sigma \times Q$ is the set of labelled transitions.*

A *behaviour* of an LTS, S, is a possibly infinite sequence alternating between states and actions $q_0\ a_1\ q_1\ a_2\ q_2\ \ldots$ where for all $i > 0$, $a_i \in \Sigma$ such that $(q_{i-1}, a_i, q_i) \in \delta$.

Let $\mathcal{B}(S)$ denote the behaviours of S starting from an initial state of S, *i.e.*, where $q_0 \in I$, and $\mathcal{B}(S, Q')$ denote behaviour of S starting in a state $q_0 \in Q'$ where $Q' \subseteq Q$. Let $st(b, i)$ denote the ith state of a behaviour b, and let $act(b, i)$ denote the *ith* action.

To facilitate reasoning about environmental interaction, we use a simple extension of LTS in which actions are partitioned into three sets: *internal* actions, *input* actions (externally observable actions controlled by the environment), and *output* actions (externally observable actions controlled by the component).

Such a partitioning of actions has been proposed for modelling reactive systems. It is central to the *I/O automata* approach of Lynch and Tuttle [19], and *interface automata* of de Alfaro and Henzinger [7]. In each of these approaches, combined automata interact by synchronising on common-named input and output actions. All automata with a given action are involved in each synchronisation on that action.

The main difference between I/O automata and interface automata is that the former are *input-enabled* meaning that input actions can never be refused. This is not the case with interface automata where the restrictions on the type of input actions and when they can occur is used to model assumptions on the system's environment.

An approach similar to interface automata has also been proposed for modelling groupware systems by Ellis [11]. This approach, inspired by Smith's work on collective intelligence in computer-based collaboration [31], has been formalised and further developed by ter Beek *et al.* [34]. The automata are referred to as *component automata*, and component automata which are formed as the composition of other component automata as *team automata*. The major difference with the aforementioned approaches to reactive systems is that in a team automaton not all of the composed automata with a given action need to synchronise on that action. This flexibility has been shown to be well suited to formalising notions of coordination, cooperation and collaboration in a distributed setting [34]. In the remainder of this section, we show how agents and MAS can be modelled using component and team automata.

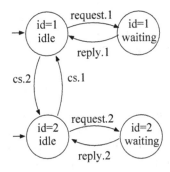

Fig. 1. Component automaton of the client agent

3.1   Agents as Component Automata

Agents are modelled as component automata [34].

Definition 2. *An agent is an LTS, $A = (Q, I, \Sigma = \Sigma_{int} \cup \Sigma_{inp} \cup \Sigma_{out}, \delta)$, where Σ_{int}, Σ_{inp} and Σ_{out} are pairwise disjoint, and*

- *Σ_{int} is the set of internal actions. Such actions are controlled by the agent and are not externally observable.*
- *Σ_{inp} is the set of input actions. Such actions are externally observable and are controlled by the agent's environment.*
- *Σ_{out} is the set of output actions. Such actions are externally observable and are controlled by the agent.*

Example 1. Consider an agent *Client* which is aware of a number of servers in its environment with which it can interact. The state of the client includes the set of server identifiers and the identifier of the server with which it is currently interacting. It has a set of internal actions *cs.id* which allows it to change the server with which it is interacting to that with identifier *id* (initially the client is interacting with any server of which the agent is aware), a set of output actions *request.id* representing a request to the server with identifier *id*, and a set of input actions *reply.id* representing a reply from the server with identifier *id*.

Assume there are two available servers with identifiers 1 and 2. An LTS that models the client can be depicted as in Fig. 1, where incoming arrows mark initial states. In this model, the client changes server only when it is not awaiting a reply.

To represent this system as a component automaton, we simply partition its actions as follows.

$$\Sigma_{int} = \{cs.i \mid i \in 1..2\}$$
$$\Sigma_{inp} = \{reply.i \mid i \in 1..2\}$$
$$\Sigma_{out} = \{request.i \mid i \in 1..2\}$$

◇

Given this partitioning, the fact that the input action *reply.id* occurs only after *request.id*, for $id \in 1..2$, is an assumption that has been made about the client's environment. It is not something the client could itself enforce.

3.2 Multi-Agent Systems as Team Automata

When component automata are composed, they potentially synchronise on common-named actions. Hence to prevent unwanted synchronisations, a precondition for composing a group of agents is that no internal action of one agent is present as an action (internal or external) of another agent.

Let A_i denote the agent $(Q_i, I_i, \Sigma_i = \Sigma_{i,int} \cup \Sigma_{i,inp} \cup \Sigma_{i,out}, \delta_i)$, for $i \in 0..n$. A composition of the agents $A_0, \ldots, A_n$ is possible if

$$\forall i \in 0..n \bullet (\Sigma_{i,int} \cap \bigcup_{j:0..n \setminus \{i\}} \Sigma_j) = \varnothing. \tag{1}$$

Given such a composable set of agents, a multi-agent system (MAS) is modelled as a special kind of component automaton called a *team automaton* [34]. A state q of the team automaton is a tuple of the possible states of the agents, $q \in \prod_{i:0..n} Q_i$. We let q_j, for $j \in 0..n$, denote the jth element of the tuple q.

Definition 3. *A MAS comprising agents* $A_0, \ldots, A_n$ *is an LTS,* $M = (Q, I, \Sigma = \Sigma_{int} \cup \Sigma_{inp} \cup \Sigma_{out}, \delta)$, *where* Σ_{int}, Σ_{inp} *and* Σ_{out} *are pairwise disjoint, and*

- $Q = \prod_{i:0..n} Q_i$.
- $I = \prod_{i:0..n} I_i$.
- $\Sigma_{int} = \bigcup_{i:0..n} \Sigma_{i,int}$.
- $\Sigma_{out} = \bigcup_{i:0..n} \Sigma_{i,out}$.
- $\Sigma_{inp} = (\bigcup_{i:0..n} \Sigma_{i,inp}) \setminus \Sigma_{out}$.
- $\delta \subseteq Q \times \Sigma \times Q$ *such that*
  - *for all* $(q, a, q') \in \delta$, *there exists a* $j \in 0..n$ *such that* $(q_j, a, q'_j) \in \delta_j$ *and for all* $i \in 0..n$ *with* $i \neq j$, $(q_i, a, q'_i) \in \delta_i$ *or* $q_i = q'_i$
  - *for all* $q, q' \in Q$ *and* $a \in \Sigma_{int}$, *if there exists a* $j \in 0..n$ *such that* $(q_j, a, q'_j) \in \delta_j$, *then* $(q, a, q') \in \delta$.

The internal and output actions of M are those of the agents. The input actions are those of the agents which are not also output actions. In the case where an input action of one agent is the same as an output action of another agent, the input is assumed to be caused by the output action and hence is not an input action for the MAS. The fact that the output action is not also removed from the system allows team automata to be further composed with other component or team automata, *e.g.*, to act as the environment of a component in a further composition.

The transitions of M are such that the following hold.

(i) Each transition involves a non-empty subset of agents engaging in the action a. The state of each agent not involved in the action remains unchanged.
(ii) There is a MAS transition for each agent transition corresponding to an internal action.

Not all agents with action a need to be involved in a system transition corresponding to a. This allows different interaction strategies to be captured [34]. However for consistency, we require that any output action of the system involves at least one agent output action, *i.e.*, there should not be a system action a involving only an agent which has a as an input action when there are other agents which have a as an output action. More formally

$$\forall (q, a, q') \in \delta \bullet a \in \Sigma_{out} \Rightarrow \exists j \in 0 \mathinner{.\,.} n \bullet a \in \Sigma_{j,out} \wedge (q_j, a, q'_j) \in \delta_j. \quad (2)$$

Furthermore, given a transition (q, a, q') of a MAS such that $q_j = q'_j$ for some $j \in 0 \mathinner{.\,.} n$, if the jth agent has a transition (q_j, a, q_j) then the agent undergoes this action, otherwise (*i.e.*, if $(q_j, a, q_j) \notin \delta_j$) it undergoes no action. This *maximal* interpretation suggested by ter Beek *et al.* [34] removes any ambiguity concerning which agents participate in a particular MAS transition.

Example 2. To continue Example 1 above we assume that each server is defined as $Server_i = (Q_i, I_i, \Sigma_{i,int} \cup \Sigma_{i,inp} \cup \Sigma_{i,out}, \delta_i)$ with $\Sigma_{i,int} = \varnothing$, $\Sigma_{i,out} = \{reply.i \mid i \in 1 \mathinner{.\,.} 2\}$ and $\Sigma_{i,inp} = \{request.i \mid i \in 1 \mathinner{.\,.} 2\}$. The behaviour of the two servers is modelled abstractly in Fig. 2. (For simplicity, we assume that a server deals with only one client at a time).

Fig. 2. Component automata of the server agents

The agents *Client*, *Server*₁ and *Server*₂ can be composed since (1) holds.

Given $Client = (Q_{Client}, I_{Client}, \Sigma_{Client}, \delta_{Client})$, one team automaton that can be composed from *Client* and the servers *Server*₁ and *Server*₂ is $M = (Q, I, \Sigma_{int} \cup \Sigma_{inp} \cup \Sigma_{out}, \delta)$ with

- $Q = Q_{Client} \times \prod_{i:1..2} Q_i$.
- $I = I_{Client} \times \prod_{i:1..2} I_i$.
- $\Sigma_{int} = \{cs.i \mid i \in 1 \mathinner{.\,.} 2\}$.
- $\Sigma_{inp} = \varnothing$.
- $\Sigma_{out} = \{reply.i, request.i \mid i \in 1 \mathinner{.\,.} 2\}$.
- $\delta = \{(q, a, q') \in Q \times \Sigma \times Q \mid (q_0, a, q'_0) \in \delta_{Client} \wedge$
  $\quad a \notin \Sigma_{int} \Rightarrow (\exists i \in 1 \mathinner{.\,.} 2 \bullet (q_i, a, q'_i) \in \delta_i \wedge (\forall j \neq i \bullet q_j = q'_j)) \wedge$
  $\quad a \in \Sigma_{int} \Rightarrow (\forall i \in 1 \mathinner{.\,.} 2 \bullet q_i = q'_i)) \}.$

Since all common-named actions synchronise, and all actions which are enabled in a component can occur, the definition satisfies (2).

It is possible, by restricting δ in such compositions, to limit when operations are enabled, or to limit the agents which synchronise on an action. For example, if we had included two client agents, then we would expect only one to be involved in each request, reply and change-server action. $\diamond$

4 Adaptivity Defined

In this section we provide formal definitions of adaptivity for agents and MAS based on their team automata representations as defined in Sect. 3. We base our definitions on Dijkstra's notion of legitimate states [8] which we extend to include both the state of the system under consideration (agent or MAS) and its environment. The definitions qualify adaptivity with respect to the external action to which the system adapts, and quantify it with respect to the number of actions required to adapt.

Since agents and MAS are represented by automata, the definition of adaptivity for each of them is identical. We begin by defining adaptivity in the special case of *closed systems*, *i.e.*, where the system does not interact with its environment, in Sect. 4.1. This definition is applicable to MAS which do not rely on environmental interaction for their operation. We then extend the definition to *open systems*, *i.e.*, where interaction with the environment is central to the system's operation, in Sect. 4.2. This definition is applicable to agents as well as MAS that interact with their environment.

4.1 Adaptivity of Closed Systems

A closed MAS M can be modelled by a team automaton with no input actions. The team automaton of Example 2 is an example of such a closed system. Let $\mathcal{Q}(M)$ denote the set of legitimate states of M. By definition, all transitions from legitimate states lead to legitimate states. That is, given $M = (Q, I, \Sigma, \delta)$

$$(q, a, q') \in \delta \wedge q \in \mathcal{Q}(M) \Rightarrow q' \in \mathcal{Q}(M). \tag{3}$$

A MAS is *well-formed* if the initial states of M are legitimate states, or if M is guaranteed to reach a legitimate state in a finite number of actions. That is,

$$I \subseteq \mathcal{Q}(M) \vee (\forall b \in \mathcal{B}(M) \bullet \exists i \geq 0 \bullet st(b, i) \in \mathcal{Q}(M)). \tag{4}$$

In the case where a finite number of actions are required to reach a legitimate state, the MAS undergoes an initial self-configuration process.[1]

Let Q be the set of states of M and Z be an external action defining the set of transitions $\zeta \subseteq Q \times Z \times Q$ on M. Such an external action can move the MAS from a legitimate state to an illegitimate one. M can adapt to the external action, if it can return to a legitimate state.

---

[1] Self-configuration can itself be viewed as a type of adaptivity in which the external action is the system initialisation.

Definition 4. *A closed MAS M is Z-adaptive if, after an occurrence of Z which places the MAS in an illegitimate state, the MAS is guaranteed to reach a legitimate state in a finite number of transitions under the assumption of no further occurrences of Z.*

That is, for all $b \in \mathcal{B}(M)$ such that there exists an $i \geq 0$ such that $st(b, i) = q$ and for all illegitimate states q' such that $(q, Z, q') \in \zeta$ the following holds.

$$\forall\, b' \in B(M, \{q'\}) \bullet \exists j > 0 \bullet st(b', j) \in \mathcal{Q}(M) \tag{5}$$

Note that we are concerned only with cases where Z places the MAS in an illegitimate state. If Z places the MAS in a legitimate state, the MAS is robust against Z, but we do not regard this as adapting (since there is no deflection from its normal behaviour).

A closed MAS M is n-Z-adaptive for some $n > 0$, if it can adapt within at most n transitions. That is, for all $b \in \mathcal{B}(M)$ such that there exists an $i \geq 0$ with $st(b, i) = q$ and for all illegitimate states q' such that $(q, Z, q') \in \zeta$ the following holds.

$$\forall\, b' \in B(M, \{q'\}) \bullet \exists j \in 1 .. n \bullet st(b', j) \in \mathcal{Q}(M) \tag{6}$$

The following theorems follow directly from these definitions.

Theorem 1. *If M is n-Z-adaptive, it is also m-Z-adaptive for any $m \geq n$.*

Theorem 2. *If M is n-Z-adaptive for some $n > 0$, then it is also Z-adaptive.*

Note that the inverse of Theorem 2 does not hold. It is possible, due to nondeterminism in a MAS, that there is no minimum number of transitions required to reach a legitimate state. For example, consider a MAS that after Z is repeatedly able to choose between two actions a and b, and reaches a legitimate state after choosing b. If we assume fairness (so that b must eventually be chosen) the MAS is Z-adaptive. However, there is no n for which it is n-Z-adaptive.

4.2   Adaptivity of Open Systems

An agent provides an example of an *open* system since its behaviour typically depends on its environment. The environment controls the agent's input actions, and may restrict the occurrence of its output actions when synchronisation is required. Similarly, a MAS can be an open system. In this section we will discuss adaptivity of agents, although the results are also directly applicable to open MAS.

To reason about an agent, we need to model the interactions with its environment. In interface automata, this is taken care of by the restrictions placed on the types of observable (input and output) actions and when they can occur [7]. The same is true for component and team automata.

With open systems, there are two kinds of external actions. The first change the state of the agent. They are identical to the external actions of closed systems and adaptivity to these actions can be reasoned about in the same way.

The second kind of external actions change the state of the system's environment, and possibly also the system's state. In the setting of component automata, this would manifest itself as (possibly) different restrictions on the observable actions.[2]

To facilitate modelling such an external action, we need to extend the agent's state with one or more auxiliary variables which the external action changes. These auxiliary variables, representing some facet of the environment, can be used to restrict when particular actions can occur. They are also used in defining the legitimate states.

Example 3. Consider the client agent of Sect. 3. A possible external action that could occur in its environment is a server going down. Let Z_1 be the external action that causes a server with which the client is interacting to go down when the client is in the state where it has not performed a request. Let Z_2 be an external action similar to Z_1 which occurs when the client has performed a request and is waiting for a reply.

To reason about the adaptivity of the client, we extend it with an auxiliary variable *down* which is the set of servers which are currently down. This set is initially empty. The transitions are restricted such that if the extended client is in a state where server i is down, transitions corresponding to *cs.i*, *request.i* and *reply.i* cannot occur. If it is in a state where server i is not down, the transitions can occur. The transitions do not change whether a given server is up or down.

Figure 3 shows part of the restricted client automaton. We show the states in which both servers are up, and also the states in which server 1 is down. We choose the legitimate states to be those where the client can perform request or reply actions, *i.e.*, $id \notin down$. These states are shaded in the figure. The external actions are shown using dotted arrows between states.

After a single occurrence of Z_1, the client is able to perform the change-server action restoring it to a legitimate state. Hence, the client is Z_1-adaptive. In fact, since it requires only one action to reach a legitimate state, it is 1-Z_1-adaptive. In a more detailed specification where, for example, the client was required to log in to the new server, the client would be n-Z_1-adaptive for some $n > 1$. Hence, the quantification of adaptivity with respect to number of actions is dependent on the level of abstraction. It should, therefore, be used only for comparing adaptive responses at the same level of abstraction.

In the case of Z_2, there is no possibility of performing the change-server action. The client is therefore not Z_2-adaptive. This may correspond to a design flaw which our reasoning allows us to detect and rectify if desired, *e.g.*, by introducing a timeout when waiting for a response.   ◇

As can be seen from Example 3, the specifier must, based on an understanding of the system and its environment, determine the available transitions from states corresponding to different values of the auxiliary variables. It is possible that

---

[2] We assume all input actions possible in the environment are included in the agent automaton, as are all of the agent's possible output actions. Hence, no observable actions will be introduced or removed by such an external action.

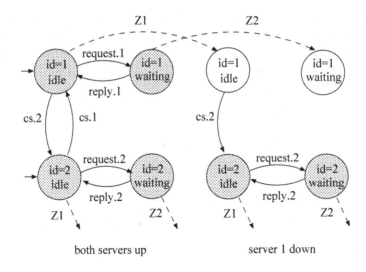

Fig. 3. Team automaton of the restricted client

these transitions change the values of the auxiliary variables. Although the agent cannot change these variables directly, it can interact with its environment to instigate their change. For example, if the agent of the above example was able to call a maintenance agent to fix the server, then as a consequence of this call it would return to a state where the server was no longer down. Whether the agent can do this and how the environment responds is up to the specifier.

To be adaptive to an external action Z, an agent must (i) be guaranteed to reach a legitimate state in a finite number of transitions, and (ii) have at least one behaviour which reaches a legitimate state without the auxiliary variables being changed. The second condition precludes agents which rely on the auxiliary variables changing to reach a legitimate state. For example, an agent may call a maintenance agent but have no other strategy for dealing with a server that is down. We would not regard such an agent as adaptive.

The approach is formalised as follows. We enhance the agent $A = (Q, I, \Sigma = \Sigma_{int} \cup \Sigma_{inp} \cup \Sigma_{out}, \delta)$ with a set of auxiliary variables describing a set of states E. The cross product $Q \times E$ captures the state space of A extended with these auxiliary variables. Thus, the enhanced version of the agent is defined as $A' = (Q \times E, I \times I_E, \Sigma, \delta')$ where $I_E \subseteq E$ and $\delta' \subseteq (Q \times E) \times \Sigma \times (Q \times E)$. We require that the behaviours of A' when restricted to Q correspond to behaviours of the original agent A. That is,

$$\forall b \in \mathcal{B}(A') \bullet b|_Q \in \mathcal{B}(A) \tag{7}$$

where $b|_Q$ denotes the behaviour b restricted to the state Q of the original agent A. Hence, A' behaves identically to A in the absence of external actions. This is true in Example 3 since initially no servers are down.

Let Z be an external action defining the set of transitions $\zeta \subseteq (Q \times E) \times Z \times (Q \times E)$ on A'.

Definition 5. *An agent (or open MAS) A is Z-adaptive, if its enhancement A'
on which Z is defined is, after an occurrence of Z which places it in an illegitimate state, able to reach a legitimate state in a finite number of transitions, and
at least one behaviour reaches a legitimate state without changing the extension
to the state of A.*

*That is, for all $b \in \mathcal{B}(A')$ such that there exists an $i \geq 0$ such that $st(b, i) = q$
and $(q, Z, q') \in \zeta$ the following holds.*

$$\forall\, b' \in \mathcal{B}(A', \{q'\}) \bullet \exists\, j > 0 \bullet st(b', j) \in \mathcal{Q}(A') \tag{8}$$

and

$$\begin{aligned} &\exists\, b' \in \mathcal{B}(A', \{q'\}) \bullet \\ &\exists\, j > 0 \bullet st(b', j) \in \mathcal{Q}(A') \wedge (\forall\, k \leq j \bullet st(b', k)|_E = q'|_E) \end{aligned} \tag{9}$$

where $s|_E$ restricts a state s of A' to the variables of E.

An agent (or open MAS) A which is Z-adaptive is n-Z-adaptive for some
$n > 0$, if it can adapt within at most n transitions. That is, for all $b \in \mathcal{B}(A')$
such that there exists an $i \geq 0$ such that $st(b, i) = q$ and $(q, Z, q') \in \zeta$, the
following holds along with condition (9) above.

$$\forall\, b' \in \mathcal{B}(A', \{q'\}) \bullet \exists\, j \in 1 .. n \bullet st(b, j) \in \mathcal{Q}(A') \tag{10}$$

Theorems 1 and 2 of Sect. 4.1 remain true and follow directly from these
definitions.

5   A Teleo-Reactive Development Framework

The notions of component and team automata introduced in Sect. 3 have provided a convenient setting for the definition of adaptivity in Sect. 4. We now
turn to the question of designing adaptive systems: specifically, to the generation of high-level (behavioural) specifications of such systems which can be
developed to suitable implementations using any of the range of techniques discussed in Sect. 2.

Our approach is to make the adaptive behaviour of the system explicit in
the specification, and thereby simplify its validation. This is achieved using a
specification framework based on the teleo-reactive agent paradigm of Nilsson
[23,24]. This paradigm, which was developed to facilitate

> "robustly [directing] an agent toward a goal in a manner that continu-
> ously takes into account the agent's changing perceptions of a dynamic
> environment" [24],

is well suited to our task.

A teleo-reactive agent is represented by an ordered list of production rules of the form

$$C \longrightarrow P$$

where C is a condition evaluated with respect to the agent's world model (comprising its perception of its own state and that of the environment), and P is a non-terminating program (referred to as a *durative action*). The behaviour of such an agent is to continuously evaluate the conditions in the list and, at any time, perform the durative action associated with the *first* production rule in the list whose condition is true. If no conditions are true, the behaviour of the teleo-reactive agent is undefined.

Definition 6. *A teleo-reactive agent is a 6-tuple $T = (Q, I, \Sigma, \delta, C, \rho)$ where*

- *(Q, I, Σ, δ) is a component automaton.*
- *$C \subseteq 2^Q$ is the set of conditions whose elements form a total order $\leq$ capturing the priority amongst production rules.*
- *$\rho \subseteq C \times 2^\delta$ is the agent's set of production rules and satisfies the following constraint. For all conditions c in C, for all $q \in c$ such that $q \notin c'$ for any $c' \leq c$, there exists a transition $(q, a, q') \in \rho(c)$. This constraint ensures the set of actions associated with a condition represents a non-terminating program, or durative action.*

Assuming transitions in δ are atomic and that conditions are evaluated before each transition, the behaviour of a teleo-reactive agent, T, is a possibly infinite sequence alternating between states and actions $q_0 \ a_1 \ q_1 \ a_2 \ q_2 \ \ldots$ where for all $i \geq 0$, if there exists a condition $c \in C$ such that $q_i \in c$ and, for all $c' \leq c$, $q_i \notin c'$ then $(q_i, a_i, q_{i+1}) \in \rho(c)$. If there does not exist a condition $c \in C$ such that $q_i \in c$ then the action and post state of the next transition are undefined; they can be any action and state including those not in the sets Q and Σ respectively. This would be used to leave implementation flexibility in an abstract design.

Teleo-reactive agents are usually designed to have a *regression* property whereby executing a durative action P_i when condition c holds, will eventually result in a condition c' occurring earlier in the list, *i.e.*, $c' \leq c$. This enables an agent to eventually reach its goal corresponding to the first production rule in the list. In our context, we use the regression property to ensure an adaptive agent reaches a legitimate state.

5.1 Designing Adaptivity

In keeping with the generality of our results, our teleo-reactive framework is independent of the specification notation used to capture the behaviour of the agent's state and actions. We illustrate its use on a case study with a particular specification language, Object-Z [28], in Sect. 6.

To motivate our approach, we return to the client-server example of Sects. 3 and 4. In Fig. 3 we extended the original component automaton for a client agent to include a modified behaviour when server 1 is down (corresponding to a particular value of the auxiliary variable *down*). Although not shown in Fig. 3, similar modified behaviours would exist for when server 2 is down and for when both servers are down. The client can therefore be thought of as being in one of four *modes*; the particular mode being determined by the state of the environment, *i.e.*, which servers are currently down.

As our definition of adaptivity precludes the agent changing its environment, reasoning about adaptivity following an external action Z is reduced to reasoning within the mode to which Z takes the agent. For example, when server 1 goes down we need consider only the behaviour within the four (out of 16) states corresponding to this server being down. We therefore define a teleo-reactive specification corresponding to each mode. This considerably simplifies the process of reasoning about adaptivity.

For the normal operational behaviour of the agent, *i.e.*, when no servers are down, the specification is as in Fig. 1. A teleo-reactive version of this specification is presented below.

$$Client \mathrel{\hat=} true \longrightarrow \{\, request, respond, cs \,\}$$

where $\{\, request, respond, cs \,\}$ denotes a program which repeatedly chooses to perform one of the agent actions *request*, *respond* or *cs*. The choice when more than one action is enabled is nondeterministic. For this program to represent a durative action, it is necessary that at least one of the agent actions is always enabled (which is the case in the example). In general, we need to prove

$$\forall c \in C \bullet \forall q \in c \bullet (\forall c' \neq c \bullet c' \leq c \Rightarrow q \notin c') \Rightarrow$$
$$\exists q' \in Q, a \in \Sigma \bullet (q, a, q') \in \rho(c). \tag{11}$$

In the teleo-reactive specification of *Client*, there are no illegitimate states and hence the teleo-reactive specification has only one production rule. In general, an agent may start in an illegitimate state and need to self-configure before normal operation begins. In that case, there would be one production rule whose condition describes the legitimate states followed by one or more production rules whose conditions describe illegitimate states. Self-configuration would be proved by showing that the specification has the previously mentioned regression property. Similarly, adaptivity to external duress can be shown in this way.

Consider the mode of the client when server 1 is down. The teleo-reactive specification is

$$Server1Down \mathrel{\hat=} id = s2 \longrightarrow \{\, request, respond \,\}$$
$$id = s1 \wedge state = idle \longrightarrow \{\, cs \,\}$$
$$id = s1 \wedge state = waiting \longrightarrow \{\, skip \,\}$$

where the special action *skip* corresponds to the agent doing nothing.

The legitimate states are those satisfying $id = s2$. We divide the illegitimate states into those that result from $Z1$ (satisfying $id = s1 \wedge state = idle$) and those that result from $Z2$ (satisfying $id = s1 \wedge state = waiting$). To prove adaptivity to $Z1$, we need to simply prove that cs changes id from $s1$ to $s2$ (since this will make the earlier condition $id = 2$ true moving the agent to a legitimate state). It is also obvious from the specification that the agent cannot adapt to $Z2$ (since $skip$ does not change the agent's state). If this were not desirable, we could change the specification at this point to add, for example, the timeout action suggested in Sect. 4.

A similar teleo-reactive specification $Server2Down$ is given for the mode in which server 2 is down. For the mode where both servers are down the teleo-reactive specification is

$$BothServersDown \mathrel{\widehat{=}} true \longrightarrow \{skip\}$$

indicating there is no means for the agent to adapt; external (maintenance) actions are required for continued operation in this case.

By considering each mode in isolation as a teleo-reactive system, we can readily reason about the adaptivity of our design and modify the design when necessary. The final step of our approach combines the mode specifications as follows.

$$
\begin{aligned}
TR\_Client \mathrel{\widehat{=}} \; & down = \varnothing \longrightarrow Client \\
& down = \{s1\} \longrightarrow Server1Down \\
& down = \{s2\} \longrightarrow Server2Down \\
& down = \{s1, s2\} \longrightarrow BothServersDown
\end{aligned}
$$

where the occurrence of a teleo-reactive specification on the right-hand side of a production rule is a syntactic convention which expands as follows: A rule $C \longrightarrow TR$, where TR is a teleo-reactive specification, is replaced by a sequence of rules $C \wedge C_i \longrightarrow P_i$ corresponding to the sequence of rules $C_i \longrightarrow P_i$ in TR. For example, $down = \{s1\} \longrightarrow Server1Down$ is replaced by

$$
\begin{aligned}
down &= \{s1\} \wedge id = s2 \longrightarrow \{request, respond\} \\
down &= \{s1\} \wedge id = s1 \wedge state = idle \longrightarrow \{cs\} \\
down &= \{s1\} \wedge id = s1 \wedge state = waiting \longrightarrow \{skip\}.
\end{aligned}
$$

Formally, a composed teleo-reactive system is defined as follows.

Definition 7. *Let a system have n modes on the state space defined by states Q and initial states I. Let $(Q, I, \Sigma_i, \delta_i, C_i, \rho_i)$ where $i \in 1 \mathinner{\ldotp\ldotp} n$ be the teleo-reactive specification of the ith mode. Let E denote the set of states described by the auxiliary variables, and $E_i \subseteq E$ be those states corresponding to the ith mode. Given the first mode corresponds to normal behaviour (in the absence of external duress), the teleo-reactive specification of the entire system is $(Q', I', \Sigma, \delta, C, \rho)$ where*

- $Q' = Q \times E$.
- $I' = I \times E_1$.
- $\Sigma = \bigcup\limits_{i:1\mathinner{\ldotp\ldotp}n} \Sigma_i.$

- $\delta = \{(q, a, q') \in Q \times \Sigma \times Q \mid \exists i \in 1 .. n \bullet q \mid_E \in E_i \wedge (q \mid_Q, a, q' \mid_Q) \in \delta_i\}$.
- $C = C_1 \times E_1 \cup \ldots \cup C_n \times E_n$ such that
  - for all $i \in 1 .. n$ and $c, c' : C_i$ where $c' \leq c$, $c' \times E_i \leq c \times E_i$
  - for all $i, j \in 1 .. n$ where $i \leq j$ and $c' : C_i$ and $c : C_j$, $c' \times E_i \leq c \times E_j$.
- $\rho \subseteq C \times 2^\delta$ such that for all $c \in C_i \times E_i$, $(q, a, q') \in \rho(c)$ iff $(q \mid_Q, a, q' \mid_Q) \in \rho_i(c \mid_Q)$.

The composed teleo-reactive system begins in the first mode and transitions to other modes depending on the values of the auxiliary variables. The behaviour in a given mode is identical to that of the mode considered in isolation. Since the first mode corresponds to the behaviour in the absence of external actions, (7) is guaranteed to hold.

It should be noted that the assumption that modes operate on the same state space is not overly restrictive. If it is necessary to specify modes of a system which operate on different states, $i.e.$, different sets of variables, to combine these modes we first extend the states of each mode with the variables of the others to provide a unified state.

5.2   From Teleo-Reactive Specification to Component Automaton

While teleo-reactive implementation approaches exist [15], in general we may not wish to implement our system in a teleo-reactive style. We therefore provide a mapping from a teleo-reactive specification to the more general form of a component automaton.

Definition 8. *Let* $(Q, I, \Sigma, \delta, C, \rho)$ *be a teleo-reactive system. The equivalent component automaton is* (Q, I, Σ, δ') *where*

$$\delta' = \bigcup_{c:C} \{(q, a, q') \in \rho(c) \mid q \in c \wedge \forall c' \leq c \bullet c' \neq c \Rightarrow q \notin c'\}.$$

A transition from the teleo-reactive system is enabled only when a condition under which it may occur is true, and all conditions of earlier production rules are false.

5.3   Adapting to Internal Disturbances

The approach proposed above assumes a set of auxiliary variables which capture the effect of an external action. In the client example these variables represent a state of the environment. The auxiliary variables can, however, also be used to represent a condition of the internal state of the system. The main difference is that it may be possible in this case for a system to move from one mode to another during adaptation. Hence, for showing that a system adapts when in a given mode may require showing that, rather than reaching a legitimate state in that mode, the system transitions to another mode from which it can adapt.

As a result the approach can be used in the development of systems (including closed systems) which adapt to changes to internal state. This is illustrated by the case study in Sect. 6.

6   Case Study: The Self-Adaptive Production Cell

In this section we illustrate our approach and its use with a specific formal notation, Object-Z [28]. Object-Z is an object-oriented extension of the well known Z specification language [32]. Its notions of classes and objects are ideal for specifying MAS [30]. Both Z and Object-Z have been advocated for the description of agents by other researchers in the field [9,14].

Our case study is a self-adaptive, multi-robot production cell based on that described by Nafz *et al.* [22]. The production cell comprises a number of robots which are capable of taking on various roles in the production of an item. These roles may involve the use of tools such as drills and screwdrivers. Since changing tools, and hence changing roles, is very time-consuming, the robots in the production cell take on different roles from each other and collaborate on the production of items; passing the items between them in the required order. For example, a typical scenario is a production cell with three robots: the first robot uses a drill to drill a hole in the item, the second inserts a screw, and the third tightens the screw with a screwdriver [22].

In the design of the system below, we consider both the self-configuration of the system from an initial state where no robot has a role, as well as the ability of the system to adapt to a robot losing a capability, *e.g.*, if a robot in the above scenario breaks its drill or screwdriver, or runs out of screws.

6.1   Normal Behaviour and Self-Configuration

We begin by specifying the normal behaviour of the system, in which all robots possess all capabilities. A robot is specified using an Object-Z class which, like a class in an object-oriented programming language, encapsulates state variables, their initial values and all operations which can change their values.

The robot has two state variables: *available* denoting the roles which are available for the robot to take on (corresponding to an internal model of its environment), and *roles* the robot's current roles (we will limit the number of roles to one in the specification, but, in general, cases where a robot could take on more than one role could be considered). Each variable is assigned a value which is a subset of a given type *Role* comprising all possible roles ($\mathbb{P}\,S$ denotes the power set of S). Initially, all roles are available and the robot has no role.

$$
\begin{array}{|l}
\hline
\_\textit{Robot}\_\!\!\rule{5cm}{0.4pt} \\
\hline
\quad available : \mathbb{P}\,Role \\
\quad roles : \mathbb{P}\,Role \\
\hline
\quad roles \cap available = \varnothing \\
\quad \#roles \leq 1 \\
\hline
\end{array}
$$

$$
\begin{array}{l}
\rule{0pt}{0pt}\textit{INIT} \\
\hline
available = Role \\
roles = \varnothing \\
\hline
\ldots \quad \text{operations detailed below}
\end{array}
$$

The operations of a class are named boxes with

- a Δ-list (read "delta list") listing the state variables which the operation may change; all other variables remain unchanged. An operation without a Δ-list cannot change any state variables.
- a number of declarations of local variables (such as inputs and outputs).
- a predicate restricting the values of the state variables both before and after the operation, and the values of the local variables. State variables after an operation are denoted by the variable name decorated with a prime, *e.g.*, *available'*.

The operations of class *Robot* are as follows.

A robot without a role may choose one from the available roles. The robot's new role is communicated to the environment via the output variable $r!$. At the level of abstraction of our specification we are assuming that two robots will not choose an available role simultaneously. At a lower level of abstraction such an occurrence would need to be resolved using a suitable contention mechanism such as the robot with the minimum (or maximum) identifier backing off, or both robots backing off for random amounts of time.

$$
\begin{array}{l}
\textit{ChooseRole} \\
\hline
\Delta(available, roles) \\
r! : Role \\
\hline
role = \varnothing \\
r! \in available \\
available' = available \setminus \{r!\} \\
roles' = \{r!\}
\end{array}
$$

A robot receiving an input $r?$ corresponding to another robot taking on an available role, updates its environmental model accordingly.

$$
\begin{array}{l}
\textit{RemoveAvailable} \\
\hline
\Delta(available) \\
r? : Role \\
\hline
r? \in available \\
available' = available \setminus \{r?\}
\end{array}
$$

A robot with a role may operate according to that role. We abstract from what the robot actually does including the passing of the item between robots: these aspects of the production cell have no effect on the system's adaptivity.

$$
\begin{array}{l}
\rule{0pt}{0pt}\underline{\quad Operate \rule{6cm}{0.4pt}} \\
\ \ roles \neq \varnothing \\
\rule{7cm}{0.4pt}
\end{array}
$$

The system is specified by another class *System* whose state comprises a set of robots; one for each role. Initially, each robot is in its initial state, *i.e.*, has no role and believes all roles are available.

$$
\begin{array}{l}
\underline{\quad System \rule{6cm}{0.4pt}} \\[4pt]
\quad robots : \mathbb{P}\,Robot \\
\quad \rule{5cm}{0.4pt} \\
\quad \#robots = \#Role \\[6pt]
\quad \underline{\quad INIT \rule{5cm}{0.4pt}} \\
\quad \ \forall\, r : robots \bullet r.INIT \\[6pt]
\quad ChooseRole \;\widehat{=}\; [\,]\ r_0 : robots \bullet r_0.ChooseAvailable \;\|\; \\
\qquad\qquad\qquad\qquad\qquad (\wedge r : robots \setminus \{r_0\} \bullet r.RemoveAvailable) \\[6pt]
\quad Operate \;\widehat{=}\; \wedge r : robots \bullet r.Operate \\
\rule{8cm}{0.4pt}
\end{array}
$$

The operations use the familiar dot notation from object-oriented programming to denote robots undergoing operations. In *ChooseRole* the choice operator $[\,]$ is used to select one robot r_0 to undergo operation *ChooseAvailable*, and the conjunction operator $\wedge$ to specify all other robots undergoing *UpdateAvailable*. The parallel composition operator $\|$ equates the inputs $r?$ of the latter robots' operations with the output $r!$ of r_0's operation. A precise semantics of these operators is given by Smith [28]. In *Operate*, the conjunction operator is used to specify all robots undergoing their *Operate* operation.

Following the approach in the previous section and assuming that legitimate states are those in which all robots have a role, the teleo-reactive specification of the system is as follows.

$$
NormalOperation \;\widehat{=}\; \forall\, r : robots \bullet r.role \neq \varnothing \;\longrightarrow\; \{Operate\} \\
\qquad\qquad\qquad\qquad true \;\longrightarrow\; \{ChooseRole\}
$$

Note that the second production rule's condition is implicitly $\exists\, r : robots \bullet r.role = \varnothing$ since this production rule is only considered when the condition of the first production rule is false.

Showing that (11) holds is straightforward. *ChooseRole* will be enabled whenever there is a robot that can undergo *ChooseAvailable*, *i.e.*, whenever there is a robot without a role (which is the implicit condition of the second production rule). Similarly, *Operate* will be enabled whenever all robots can undergo

Operate, *i.e.*, whenever all robots have a role (which is the condition of the first production rule).

To prove the system self-configures, we need to show that the teleo-reactive specification has the regression property, *i.e.*, that the condition of the first production rule will become true. Choosing the number of robots without a role as a variant, we can show that this variant will decrease by one with each occurrence of *ChooseRole*. Hence, with a finite number of robots the variant will decrease to zero and the condition of the first production rule will be true.

6.2   Adapting to Loss of Capabilities

We now consider two external actions that cause a single robot to lose a capability. The first action $Z1$ causes it to lose the capability to perform its current role. The second $Z2$ causes it to lose the capability to perform a role other than its current role, if any. We assume that legitimate states are those in which all robots are assigned a role that they can perform.

To model the effects of these external actions, we introduce an auxiliary variable *capabilities* : $Robot \nrightarrow \mathbb{P} \, Role$ which maps robots in the system to those roles they are capable of performing. Given this variable, we will consider two modes of operation: the mode where all robots can perform all roles $\forall \, r : robots \bullet capabilities(r) = Role$, and the mode where a robot $r_0 : robots$ has lost a capability $c : Role$, $capabilities(r_0) = Role \setminus \{c\} \wedge (\forall \, r : robots \setminus \{r_0\} \bullet capabilities(r) = Role)$.

The behaviour of the former mode is captured by the teleo-reactive specification *NormalOperation* above. The behaviour of the latter is captured by the following teleo-reactive specification.

$$Loss(r_0, c) \mathrel{\widehat{=}} r_0.role \neq \{c\} \longrightarrow NormalOperation$$
$$true \longrightarrow \{ ChooseRole, skip \}$$

That is, when r_0's role is not c the system behaves as in mode *NormalOperation*, and when r_0's role is c the system may perform *ChooseRole* (when robots other than r_0 need to choose a role) and otherwise does nothing, *i.e.*, *skip*. Again it is straightforward to show (11) holds.

It is immediately obvious from the final production rule that the system is not Z_1-adaptive. To be able to adapt, it would need a way for the robot r_0 to release its current role in order to choose a new role; something we have not included in our specification. Further analysis shows that the system is also not $Z2$-adaptive: when r_0 has no role, it may choose c causing the condition of the final production rule to be the only one enabled.

Modifying the Design. In the interests of making the production cell adaptive, we modify the original specification as follows. Firstly, we add a variable *capable_of* : $\mathbb{P} \, Role$ to the class *Robot* to make robots self-aware of their own capabilities. Initially, this variable would be assigned the value *Role* and no operation would change this value (it would only be changed by external actions such as $Z1$ and $Z2$). The operation *ChooseRole* would be modified so that a robot would only choose a role it was capable of performing, *i.e.*, we would add the predicate $r! \in capable\_of$.

We would also add a new operation *ReleaseRole* which allows a robot to release a role it is not capable of performing.

$$
\begin{array}{|l}
\hline
\ \underline{\ ReleaseRole}\ \underline{\hspace{6cm}} \\
\ \Delta(available, role) \\
\ r! : Role \\
\hline
\ role = \{r!\} \\
\ r! \notin capable\_of \\
\ available' = available \cup \{r!\} \\
\ role' = \varnothing \\
\hline
\end{array}
$$

Other robots would need to update their beliefs about the available roles using the following operations.

$$
\begin{array}{|l}
\hline
\ \underline{\ AddAvailable}\ \underline{\hspace{6cm}} \\
\ \Delta(available) \\
\ r? : Role \\
\hline
\ r? \notin available \\
\ available' = available \cup \{r?\} \\
\hline
\end{array}
$$

The corresponding operation added to class *System* would be

$$ReleaseRole \mathrel{\widehat{=}} [] \ r_0 : robots \bullet r_0.ReleaseRole \ \| $$
$$ (\wedge r : robots \setminus \{r_0\} \bullet r.AddAvailable) $$

Before reanalysing the mode $Loss(r_0, c)$ with this new specification, we first re-examine *NormalOperation* to make sure that the changes have not affected the system's ability to self-configure. *NormalOperation* is in fact unchanged since, in the absence of external actions, $capable\_of = Role$ for all robots, and *ReleaseRole* is never enabled.

A naive reinterpretation of mode $Loss(r_0, c)$ is

$$Loss(r_0, c) \mathrel{\widehat{=}} r_0.role \neq \{c\} \longrightarrow NormalOperation$$
$$true \longrightarrow \{ChooseRole, ReleaseRole\}$$

However, it is easy to show that (11) no longer holds. When r_0 is the only robot without a role, and c is the only available role then *ChooseRole* is not enabled. Hence, $Loss(r_0, c)$ needs to be modified to

$$Loss(r_0, c) \mathrel{\widehat{=}} \forall \, r : robots \bullet r.role \neq \varnothing \longrightarrow \{Operate\}$$
$$r_0.role \neq \{c\} \longrightarrow \{ChooseRole, skip\}$$
$$true \longrightarrow \{ChooseRole, ReleaseRole\}$$

from which it is easy to see that adaptivity is not assured: since *skip* is the only action enabled when r_0 is the only robot without a role, and c is the only available role.

Modifying the Design Further. To remedy the above situation, we need robots to be able to release their current role for another robot to take on. This should only occur when the latter robot has lost the capability of a role and no other roles are available.

We add a boolean variable *reconfig* to class *Robot* to denote when a robot without a role is unable to choose any available role. This variable acts as a shared flag indicating that (partial) reconfiguration is required. This variable would initially be false. An operation *InitiateReconfig* sets it to true when a robot without a role cannot choose an available role.

$$
\begin{array}{l}
\underline{\ InitiateReconfig}\ \underline{\hspace{7cm}} \\
\Delta(reconfig) \\
\hline
role = \varnothing \\
available \cap capable\_of = \varnothing \\
reconfig' \\
\end{array}
$$

Other robots would need to update their *reconfig* variable using the following operation.

$$
\begin{array}{l}
\underline{\ SetReconfig}\ \underline{\hspace{7cm}} \\
\Delta(reconfig) \\
\hline
reconfig' \\
\end{array}
$$

The corresponding operation added to class *System* would be

$$
InitiateReconfig \;\widehat{=}\; [] \; r_0 : robots \bullet r_0.InitiateReconfig \wedge
$$
$$
(\wedge r : robots \setminus \{r_0\} \bullet r.SetReconfig)
$$

Note that conjunction is used in place of parallel composition here as there are no input and output variables in the combined operations.

A further operation *ChangeRole* is added to *Robot*. This operation is enabled when *reconfig* is true and the robot is capable of performing a role that is available. The robot releases its current role and takes on an available role. It also sets its *reconfig* value to false.

$$
\begin{array}{l}
\underline{\ ChangeRole}\ \underline{\hspace{7cm}} \\
\Delta(available, role, reconfig) \\
r\_old! : Role \\
r\_new! : Role \\
\hline
reconfig \\
r\_new! \in available \cap capable\_of \\
role = \{r\_old!\} \\
available' = (available \cup \{r\_old!\}) \setminus \{r\_new!\} \\
role' = \{r\_new!\} \\
\neg\ reconfig' \\
\end{array}
$$

Other robots would need to update their beliefs about the available roles and *reconfig* using the following operation.

$$
\begin{array}{|l}
\hline
\textit{ChangeAvailable} \\\hline
\Delta(\textit{available}, \textit{reconfig}) \\
r\_old? : Role \\
r\_new? : Role \\\hline
\textit{reconfig} \\
r\_old? \notin \textit{available} \\
r\_new? \in \textit{available} \\
\textit{available}' = (\textit{available} \cup \{r\_old!\}) \setminus \{r\_new!\} \\
\neg \, \textit{reconfig}' \\\hline
\end{array}
$$

The corresponding operation added to class *System* would be

$$ChangeRole \mathrel{\widehat{=}} [\!] \; r_0 : robots \bullet r_0.ChangeRole \; \|$$
$$(\wedge r : robots \setminus \{r_0\} \bullet r.ChangeAvailable)$$

Again we need to re-examine the mode *NormalOperation* in light of the changes. As the new operations become enabled only when a robot loses a capability, there is no change to behaviour. Next we need to reinterpret mode $Loss(r_0, c)$ as a teleo-reactive system. One possible specification is

$$Loss(r_0, c) \mathrel{\widehat{=}} r_0.role \neq \{c\} \wedge r_0.role \neq \varnothing \longrightarrow NormalOperation$$
$$r_0.role = \varnothing \longrightarrow Reconfigure(r_0)$$
$$r_0.role = \{c\} \longrightarrow \{ChooseRole, ReleaseRole\}$$

$$Reconfigure(r_0) \mathrel{\widehat{=}} r_0.available \cap r_0.capable\_of \neq \varnothing \longrightarrow \{ChooseRole\}$$
$$r_0.reconfig \longrightarrow \{ChooseRole, ChangeRole\}$$
$$true \longrightarrow \{ChooseRole, InitiateReconfig\}$$

It is now possible to show that (11) holds: each of the conditions ensures that the associated atomic actions form a durative action. It is also possible to show that the system is both Z_1-adaptive and Z_2-adaptive. We may further quantify the adaptivity.

Z_1 causes a robot r_0 to lose the capability to perform its current role. Hence, it will need to perform *ReleaseRole*. The worst case, in terms of number of actions until it is in a legitimate state, occurs when all other robots have no role, and then choose all roles apart from the one r_0 has just released. This would force r_0 to perform *InitiateReconfig* after which another robot would perform *ChangeRole* before r_0 could finally choose a role. In this scenario each robot in the system performs *ChooseRole*, r_0 additionally performs *ReleaseRole* and *InitiateReconfig*, and another robot performs *ChangeRole*. Hence, for n robots the system is $(n + 3)$-$Z1$-adaptive.

Z_2 causes a robot r_0 to lose a capability other than its current role, if any. The worst case occurs when all robots, including r_0, have no roles and then all robots other than r_0 choose a role other than the one r_0 has just released. Again r_0 would be forced to perform *InitiateReconfig* and another robot would have to perform *ChangeRole* before r_0 could choose a role. Hence, for n robots the system is $(n + 2)$-$Z2$-adaptive.

Examining worst case scenarios in order to quantify adaptivity can highlight inefficiencies in the design. Above it can be seen that if the robots were aware of which role had been released, they could choose that role with priority thereby avoiding a later reconfiguration. At this stage the designer may make a decision to change the design to reflect this.

Further external actions and associated modes can then be considered in a similar fashion. For example, we might like to consider where a robot r_0 loses more than one capability. In this case, the boolean variable *reconfig* is not enough: robots changing their role cannot be certain that the role they are giving up is one which r_0 is capable of performing. Hence, *reconfig* would need to be replaced by a set of roles that r_0 cannot perform, for example. We could also consider multiple robots losing capabilities, or a single robot losing all capabilities. In the latter case, for the system to adapt robots would need to be able to take on more than one role simultaneously. This would allow the system to keep functioning but, due to the time needed to swap tools, in a degraded fashion. If multiple robots could fail completely then to reduce this degradation in performance, the functioning robots would need to share the load as evenly as possible. Our approach helps us to realise these requirements and prove that suitable collaborative mechanisms satisfy them.

6.3   Combining the Modes

With just the two modes we have considered, the teleo-reactive specification of the production cell would be

$$
\begin{aligned}
ProductionCell \;\widehat{=}\; & \\
\forall\, r : robots &\bullet capabilities(r) = Role \longrightarrow NormalOperation \\
\exists\, r_0 : robots;\; & c : Role \bullet capabilities(r_0) = Role \setminus \{c\} \land \\
& (\forall\, r : robots \setminus \{r_0\} \bullet capabilities(r) = Role) \longrightarrow Loss(r_0, c)
\end{aligned}
$$

It is straightforward to show that the equivalent component automaton has the same behaviour as the modified Object-Z specification. In this case study, the effect of the auxiliary variable *capabilities* to restrict actions in a given mode is captured by the introduced variable *capable_of* of the individual robots, and so does not need to appear in the final specification.

7   Conclusion

In this paper we have presented a formal definition of adaptivity in terms of Dijskstra's notion of self stabilisation [8], and based on this definition, a framework for designing adaptive systems. The latter is based on Nilsson's notion of teleo-reactive agents [23,24]. Both our definition and framework are independent of any specific implementation mechanism for adaptivity, and any specific specification notation. We illustrated our approach using the Object-Z specification notation [28] on a case study involving a multi-robot production cell [22].

Formalisms by which adaptivity can be specified and verified are scant. Anceaume *et al.* [1] concentrate on self-organisation in dynamic networks. They argue that a definition of self-organisation cast in terms of convergence to predefined legitimate states is inadequate for two reasons: (a) the identification of legitimacy may be impractical (it may be emergent; the status of batteries in sensor networks is quoted as an example); and (b) due to the dynamic nature of the network, convergence to a legitimate state may not be possible. As a result they consider instead structural properties reflecting high interactivity between nodes, node mobility and heterogeneity, and view local self-organisation as reducing system entropy. Criticisms (a) and (b) may well be valid if only 'static' system descriptions are considered but are overcome if the dynamic features of the system are specified. Generally that requires non-local specifications; but a specification is not constrained to be local like the implementation. A simple but typical example is the glider in Conway's *Game of Life* where global time may be introduced as a 'specification device' to capture a property which emerges in the implementation from local interactions [27].

By pursuing that approach we have here offered a formalism complementary to that of [1], able to consider directly the system functionality of entirely general systems rather than secondary structural properties of dynamic networks.

An alternative to our approach should be mentioned, in which the specification of an adaptive MAS is not static but thought to change with time, reflecting adaptivity. That is the approach considered by Artikis [2] who defines a dynamic specification language C+, a kind of action language for expressing changing properties, and a 'causal calculator' for their execution. Our view is that a specification which by its definition changes with time offers none of the advantages expected of a specification. Much of the detail incurred by the present work solves the problem of how to capture, in a static specification, dynamic changes which are environmentally triggered.

Acknowledgments. This work was supported by Australian Research Council (ARC) Discovery Grant DP110101211 and the Macao Science and Technology Development Fund under the EAE project, grant number 072/2009/A3.

References

1. Anceaume, E., Défago, X., Potop-Butucaru, M., Roy, M.: A framework for proving the self-organization of dynamic systems. CoRR, abs/1011.2312 (2010)
2. Artikis, A.: A formal specification of dynamic protocols for open agent systems. CoRR, abs/1005.4815 (2010)
3. Böcker, J., Schulz, B., Knoke, T., Fröhleke, N.: Self-optimization as a framework for advanced control systems. In: Industrial Electronics Conference (IECON 2006), pp. 4671–4675. IEEE (2006)

4. Bruni, R., Corradini, A., Gadducci, F., Lluch Lafuente, A., Vandin, A.: A conceptual framework for adaptation. In: de Lara, J., Zisman, A. (eds.) FASE 2012. LNCS, vol. 7212, pp. 240–254. Springer, Heidelberg (2012)
5. Bucchiarone, A., Lafuente, A.L., Marconi, A., Pistore, M.: A formalisation of adaptable pervasive flows. In: Laneve, C., Su, J. (eds.) WS-FM 2009. LNCS, vol. 6194, pp. 61–75. Springer, Heidelberg (2010)
6. Dastani, M., Hindriks, K.V., Meyer, J.-J.C. (eds.): Specification and Verification of Multi-agent Systems. Springer, Heidelberg (2010)
7. de Alfaro, L., Henzinger, T.A.: Interface automata. In: Symposium on Foundations of Software Engineering, pp. 109–120. ACM Press (2001)
8. Dijkstra, E.W.: Self-stabilizing systems in spite of distributed control. Commun. ACM **17**, 643–644 (1974)
9. d'Inverno, M., Luck, M.: Development and application of a formal agent framework. In: International Conference on Formal Engineering Methods (ICFEM'97), pp. 222–231. IEEE Press (1997)
10. Eiben, A.E., Smith, J.E.: Introduction to Evolutionary Computing. Natural Computing Series. Springer, Heidelberg (2003)
11. Ellis, C.A.: Team automata for groupware systems. In: Hayne, S., Prinz, W. (eds.) International ACM SIGGROUP Conference on Supporting Group Work: The Integration Challenge, pp. 415–424. ACM Press (1997)
12. Georgiadis, I., Magee, J., Kramer. J.: Self-organising software architectures for distributed systems. In: Workshop on Self-healing Systems (WOSS '02), pp. 33–38 (2002)
13. Gouda, M.G., Herman, T.: Adaptive programming. IEEE Trans. Softw. Eng. **17**(9), 911–921 (1991)
14. Gruer, P., Hilaire, V., Koukam, A., Cetnarowicz, K.: A formal framework for multi-agent systems analysis and design. Expert Syst. Appl. **23**(4), 349–355 (2002)
15. Gubisch, G., Steinbauer, G., Weiglhofer, M., Wotawa, F.: A teleo-reactive architecture for fast, reactive and robust control of mobile robots. In: Nguyen, N.T., Borzemski, L., Grzech, A., Ali, M. (eds.) IEA/AIE 2008. LNCS (LNAI), vol. 5027, pp. 541–550. Springer, Heidelberg (2008)
16. Güdemann, M., Nafz, F., Ortmeier, F., Seebach, H., Reif, W.: A specification and construction paradigm for organic computing systems. In: IEEE International Conference on Self-Adaptive and Self-Organizing Systems (SASO 2008), pp. 233–242. IEEE Computer Society Press (2008)
17. Hölzl, M., Wirsing, M.: Towards a system model for ensembles. In: Agha, G., Danvy, O., Meseguer, J. (eds.) Formal Modeling: Actors, Open Systems, Biological Systems. LNCS, vol. 7000, pp. 241–261. Springer, Heidelberg (2011)
18. Hunter, A., Delgrande, J.P.: Iterated belief change: a transition system approach. In: International Joint Conference on Artificial Intelligence (IJCAI05), pp. 460–465 (2005)
19. Lynch, N., Tuttle, M.: An introduction to Input/Output automata. CWI Q. **2**(3), 219–246 (1989)
20. Mitchell, T.: Machine Learning. McGraw Hill, New York (1997)
21. Mohyeldin, E., Fahrmair, M., Sitou, W., Spanfelner, B.: A generic framework for context aware and adaptation behaviour of reconfigurable systems. In: IEEE International Symposium on Personal Indoor and Mobile Radio Communications (PIMRC05). IEEE Press (2005)

22. Nafz, F., Ortmeier, F., Seebach, H., Steghöfer, J.-P., Reif, W.: A universal self-organization mechanism for role-based organic computing systems. In: González Nieto, J., Reif, W., Wang, G., Indulska, J. (eds.) ATC 2009. LNCS, vol. 5586, pp. 17–31. Springer, Heidelberg (2009)
23. Nilsson, N.: Teleo-reactive programs for agent control. J. Artif. Intell. Res. 1, 139–158 (1994)
24. Nilsson, N.: Teleo-reactive programs and the triple-tower architecture. Electron. Trans. Artif. Intell. 5, 99–110 (2001)
25. Polani, D.: Foundations and formalizations of self-organization. In: Prokopenko, M. (ed.) Advances in Applied Self-organizing Systems. Advanced Information and Knowledge Processing, pp. 19–37. Springer, Heidelberg (2008)
26. Sanders, J.W., Smith, G.: Assuring adaptive behaviour in self-organising systems. In: Self-Organising and Self-Adaptive Systems Workshop (SASOW 2010), pp. 172–177. IEEE Computer Society Press (2010)
27. Sanders, J.W., Smith, G.: Emergence and refinement. Formal Aspects Comput. 24(1), 45–65 (2012)
28. Smith, G.: The Object-Z Specification Language. Kluwer, Norwell (2000)
29. Smith, G., Sanders, J.W., Winter, K.: Reasoning about adaptivity of agents and multi-agent systems. In: International Conference on Engineering of Complex Computer Systems (ICECCS 2012). IEEE Computer Society Press (2012)
30. Smith, G., Winter, K.: Incremental development of multi-agent systems in Object-Z. In: Software Engineering Workshop (SEW-35). IEEE Computer Society Press (2012)
31. Smith, J.B.: Collective Intelligence in Computer-Based Collaboration. Lawrence Erlbaum Associates, Hillsdale (1994)
32. Spivey, J.M.: The Z Notation: A Reference Manual, 2nd edn. Prentice-Hall International, Englewood Cliffs (1992)
33. Sutton, R.S., Barto, A.G.: Reinforcement Learning: An Introduction. MIT Press, Cambridge (1998)
34. ter Beek, M., Ellis, C., Kleijn, J., Rozenberg, G.: Synchronizations in team automata for groupware systems. Comput. Support. Coop. Work: J. Collab. Comput. 12(1), 21–69 (2003)
35. Valiant, L.: A theory of the learnable. Commun. ACM 27, 1134–1142 (1984)
36. Wooldridge, M.: An Introduction to Multiagent Systems. Wiley, New York (2002)
37. Zambonelli, F., Omicini, A.: Challenges and research directions in agent-oriented software engineering. Auton. Agent. Multi-Agent Syst. 9(3), 253–283 (2004)

Towards Formal Modelling and Verification of Pervasive Computing Systems

Yan Liu[1]([⊠]), Xian Zhang[1], Yang Liu[2], Jin Song Dong[1], Jun Sun[3], Jit Biswas[4], and Mounir Mokhtari[5]

[1] School of Computing, National University of Singapore, Singapore, Singapore
{yanliu,zhangxi5,dongjs}@comp.nus.edu.sg
[2] School of Computer Engineering,
Nanyang Technological University, Singapore, Singapore
yangliu@ntu.edu.sg
[3] Singapore University of Technology and Design, Singapore, Singapore
sunjun@sutd.edu.sg
[4] Networking Protocols Department,
Institute for Infocomm Research, Singapore, Singapore
biswas@i2r.a-star.edu.sg
[5] CNRS-IPAL, Institut TELECOM, Paris, France
Mounir.Mokhtari@it-sudparis.eu

Abstract. Smart systems equipped with emerging pervasive computing technologies enable people with limitations to live in their homes independently. However, lack of guarantees for correctness prevent such system to be widely used. Analysing the system with regard to correctness requirements is a challenging task due to the complexity of the system and its various unpredictable faults. In this work, we propose to use formal methods to analyse pervasive computing (PvC) systems. Firstly, a formal modelling framework is proposed to cover the main characteristics of such systems (e.g., context-awareness, concurrent communications, layered architectures). Secondly, we identify the safety requirements (e.g., free of deadlocks and conflicts) and specify them as safety and liveness properties. Furthermore, based on the modelling framework, we propose an approach of verifying reasoning rules which are used in the middleware for perceiving the environment and making adaptation decisions. Finally, we demonstrate our ideas using a case study of a smart healthcare system. Experimental results show the usefulness of our approach in exploring system behaviours and revealing system design flaws such as information inconsistency and conflicting reminder services.

1 Introduction

Pervasive computing(PvC) aims to provide people with a more natural way to interact with information and services by embedding computation into the environment as unobtrusively as possible [1,2]. With the rapid increase of ageing population in all industrialised societies which raised serious problems, smart

© Springer-Verlag Berlin Heidelberg 2014
R. Kowalczyk and N.T. Nguyen (Eds.): TCCI XVI, LNCS 8780, pp. 62–91, 2014.
DOI: 10.1007/978-3-662-44871-7_3

healthcare systems equipped with PvC technology are greatly needed to assist the independent living of elderly people. Such systems make it possible for elderly people to stay in their homes longer and manage everyday tasks without significant burden for their caregivers [3]. PvC systems are context-aware and adaptable to the evolving environment [4]. The changes in the environment are monitored and recorded in the system as contexts. If a particular event happens, the system is able to adapt itself to the changes. As shown in Fig. 1, a typical PvC system usually involves sensors to monitor environment changes, a residential getaway to translate the raw sensor data to low-level contexts, an inference engine to aggregate these contexts and perform user activities recognition and a reminder service to render a proper reminder for a proper user. Consequently, the heterogeneity of technology and massive ad hoc interactions among layers make PvC systems highly complicated [5]. Additionally, various environment inputs and unpredictable user behaviours cause the system behaviours beyond control, especially when multiple users are interacting with the system simultaneously.

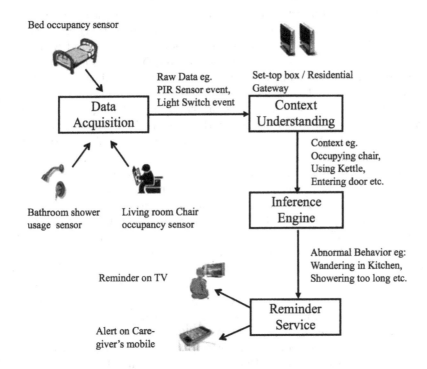

Fig. 1. A typical PvC system

Therefore, it is a challenging task to guarantee the correctness of such systems. Traditional validation methods such as simulation and testing have their limitations in performing this task. By nature, these methods only cover partial system behaviours based on the selected scenarios. Nevertheless, it is costly and time consuming to perform testing on practical systems which may require all sensors to be deployed and normal persons to act like real users. Furthermore, it is impossible to generate all kinds of environment inputs and simulate the user behaviours, especially when more than one users are interacting with the system. Last but not least, if an error is found, it is very hard to pinpoint the source since faults could appear in multiple layers such as the hardware fails to response, a reasoning rule is falsely defined or a bug is in the software system.

In order to overcome these limitations, we propose to use formal methods in the early design stage to analyse PvC systems. By a proper abstraction of the system, we are able to formally describe the system behaviours and user actions using current modelling constructs. Based on the modelling, the correctness and safety requirements can be formally specified as properties that are verifiable against the model. Existing verification techniques are reused to validate the properties by an exhaustive search of the complete system states. Counterexamples are generated to give clues for debugging. The contributions of our work are four-folds as explained below.

Firstly, we propose a framework to formally model the system design and the environment inputs. Important characteristics of PvC systems such as context-awareness, layered architecture and concurrent communications are discussed. Modelling patterns for these features are provided and illustrated with examples. In this work, we adopt CSP# [6] as the exemplar modelling language for its rich set of syntax and many extensions. Dong et al. [7] and Coronato et al. [8] proposed to model such systems using TCOZ [9] and Ambient Calculus [10] respectively. Although these languages are good at modelling the communications and mobility features respectively, the support for modelling hierarchical structures is limited. Most importantly, there is very little tool support for these languages, which limits the usage and applicability of their approaches.

Secondly, we propose critical properties and their specification with regards to the correctness requirements from stakeholders (system designers and end users). In the state of art, Arapinis et al. in [11] proposed some critical requirements of a homecare system. For instance, "Sensors are never offline when a patient is in danger" or "If a patient is in danger, assistance should arrive within a given time". In our work, we classify the critical requirements into safety properties (nothing *bad* happens) and liveness properties (something *good* eventually happens). Furthermore, formal specification patterns of these properties are proposed. As a result, we can verify the critical properties against the system design model by using automatic verification techniques like model checking [12]. Hence, design flaws can be detected at the early design stage.

Thirdly, based on the modelling framework, we propose an approach of automatic rules verification. Rules in PvC systems play a critical role. They are used to aggregate and reason the context data and decide on the adaptation decisions. As shown in Fig. 1, based on the contexts that person in the shower room, using shower tap and tap on for 30 min, the abnormal behaviour, "showering too long" is recognised. The rule defined for this behaviour is triggered and the adaptation decision is to prompt a reminder asking the person to stop showering. Generally speaking, these rules decide the responsive behaviours of the system to the users. Thus, it is essential to assure the correctness of these rules. In the literature, there is very limited work, e.g., [13,14] on rules verification in PvC domain. Most of these existing works are not directly applicable to our system. It is because they are limited to syntactic checking of relations between rules. The contribution of our work lies in redefining rule anomalies based on their execution behaviours and detecting these anomalies with the changing knowledge base.

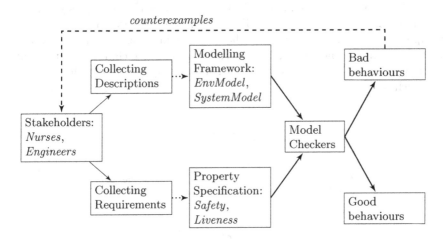

Fig. 2. Formal analysis workflow

Finally, we demonstrate the usefulness of our approach using a case study of a smart healthcare system for mild dementia patients, AMUPADH [15]. A typical workflow of this formal analysis process is shown in Fig. 2. We start the project with collecting requirements through multiple visits to the nursing home and interviews of nurses/doctors. From discussions with system designers, we learn that AMUPADH is a typical PvC system which incorporates sensors and a reasoning engine to understand the patients' intentions and provides reminder services to help them. Additionally, AMUPADH has a multi-person sharing environment which exhibits additional complexity in terms of concurrent interactions. Then, we model the user behaviours and system design based on our modelling framework using CSP# language. Critical properties such as deadlock freeness, guaranteed reminder service and conflicting reminders tests are verified using PAT model checker [16] Multiple unexpected bugs such as information inconsistency are exposed.

Rest of the paper are organised as Sect. 2 introduces the motivating example AMUPADH system; Sects. 3 and 4 demonstrates our modelling framework of PvC systems and specifications of critical requirements for verification; Sect. 5 presents the rules verification using our modelling framework. Case study on AMUPADH comes in Sect. 6. Section 7 discusses the related work while Sect. 8 concludes the paper with future directions.

2 A Motivating Example: AMUPADH - An Ambient Assisted Living System for Dementia Healthcare

Dementia is a progressive, disabling, chronic disease common in elderly people. Elders with dementia often have declining short-term memory and have difficulties in remembering necessary activities of daily living (ADLs). However, they are able to live independently or in assisted living facilities with little supervision. Ambient Assisted Living (AAL) systems equip the environment with a spectrum of computation and communication devices that seamlessly augment human thoughts and activities. AMUPADH is an AAL system deployed in a Singapore Based nursing home, Peaceheaven Nursing Home[1]. It is able to monitor the patients' behaviours using activity recognition techniques (sensors and reasoning rules) and offer help to the patients (prompt reminders through actuators such as speakers etc.).

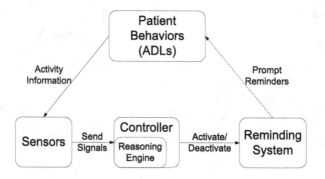

Fig. 3. An overview of the smart bedroom system

2.1 System Overview

The architecture of the system is shown in Fig. 3. The system is deployed in a room with two beds and a shower room. Different kinds of sensors are deployed to capture environment changes. For instance, the pressure sensor under a mattress is used to detect whether the bed is empty or occupied. Sensors communicate with the middleware via wireless network. The *controller* in the middleware translates sensor signals into low-level contexts from which high-level contexts are inferred by the *reasoning engine*. This reasoning task is performed based on a set of predefined rules written in Drools[2] which is a rule language based on First Order Logic. Evaluation of these rules is triggered by a sensor message or periodically by a timer. In the case that a rule is satisfied, the system will adapt to a new state by updating internal variables or invoking reminder services. For example, if the activity of patient sleeping on a wrong bed is recognised, the system will prompt a reminder requesting him to use his own bed.

---

[1] Located at 9 Upper Changi Road North, Singapore, 507706. Tel: +65-65465678.
[2] Drools Expert: http://www.jboss.org/drools/drools-expert.html

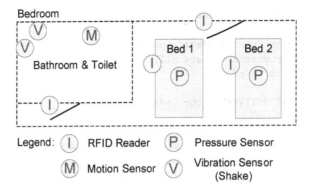

Fig. 4. Sensor layout in the bedroom

2.2 Sensors

In AMUPADH, four types of sensors are deployed in the bedroom and shower room to monitor the activity of dementia patients as shown in Fig. 4.

- **RFID Reader** is for identification and tracking. There are two readers placed beside the doors to detect who has entered the rooms respectively and two attached to each bed to identify who is using the bed. Each patient is wearing an RFID tag placed in a wrist band.
- **Pressure Sensor** is placed under the mattress of each bed to detect activities in bed, e.g., sitting or lying.
- **Shake Sensor** can detect vibration. They are attached to water pipe and soap dispenser for sensing the usage of water tap and soap respectively.
- **Motion Sensor** (A.K.A. passive infrared sensor (PIR)) can measure infrared light radiating from objects in its range. It is used to detect the presence of the patient in the shower room.

2.3 Controller

In the *Controller*, contexts are managed and inferenced. It has two components i.e., the *Main Interface* interprets the sensor signals and triggers the evaluation of all rules when a sensor message arrives; the *Context Checker* evaluates all rules every 5 min. The value is carefully tuned by system engineers in consideration of energy saving and slow user movements in AMUPADH system. The context checker is an additional step to make sure the consistency of contexts. The rules are written in Drools and evaluated by the business rule engine, Drools Expert. They are specified with a name, a condition formed of predicates and the adaptation actions. For example, the rule for detecting sitting bed for too long is specified as follows.

```
rule "personA sat on Bed A for too long (30mins)"
  when
    Sensor( id == "pressureBedA", pressureState ==
          Sensor.pressure_state.SITTING, duration > 30 )
    $x : XMPPInterface()
  then
    $x.SendData("ACTIVITY.error."+"SitBedTooLong"+"." +"personA");
end
```

The condition of this rule consists of three context variables: the sensor's *id*, status and timer. This rule can be interpreted as: the message *ACTIVITY.error. SitBedTooLong.personA* will be delivered to the reminding system if the *SITTING* status of pressure sensor on bed A has lasted for more than 30 min. The messages are sent out via a shared bus. The full set of 23 rules used in the system is listed in [17].

2.4 Reminding System

The reminding system in the application layer activates/deactivates reminders based on the incoming messages from *Controller*. For example, if the message is *ACTIVITY.error.SitBedTooLong.personA*, the reminding system decodes it and knows patient A (named Jim) has sleeping problems. Thus it invokes a speaker and prompts *'Jim, you have been sitting on bed for a long time, please go to sleep'*. This reminder will be continuously repeated until proper actions have been taken. If the prompts reach the maximum number, an alert will be sent to nurses.

3   A Modelling Framework for PvC Systems

The general picture of a PvC system is shown in Fig. 5. The system adopts a layered design and seamless interacts with the environment. Modelling of a PvC systems involves not only the modelling of important features of each important component but also the modelling of the environment inputs which play an important role in PvC systems but is often ignored in most system models.

3.1   Modelling Environments

PvC systems seamlessly interact with the environments and acquire context inputs from the users and objects like TVs and Beds. PvC systems are often driven by the environment context change (we call it *scenario* here). For example, a person entering an empty room will trigger the lights to be switched on; or when the system detects the time is 9:00pm, a take-medicine-reminder will be sent to the patient. Thus, it is important to model the scenarios with the system design. Meanwhile, the scenario model is also important for generating meaningful counterexamples so as to alleviate the burden of analysing verification results.

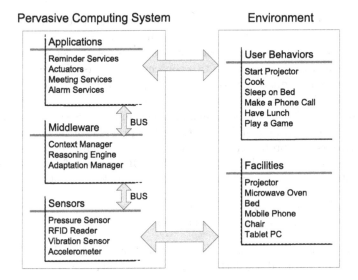

Fig. 5. Architectures of PvC systems

Modelling Activities and Environment Objects. User behaviours are various and usually unpredictable. For most PvC systems, we can observe that: (1) the system usually targets a certain group of activities and ignores other irrelevant ones; (2) relevant user activities are determined but the order of them is unpredictable. For instance, after entering a room, a person may directly go to sleep or he could possibly enter the shower room for other activities. In practice, targeted activities can be provided by system designers. We use a shower room scenario to demonstrate the modelling patterns.

In the shower room, a user performs many activities such as wandering or turning on the shower tap. These activities can be modelled as *events* which are abstractions of the observations. For example, an activity represented as event *exitShowerRoom* is an observation of the user's behaviour of leaving the shower room. However, it requires more advanced language constructs such as non-deterministic choices to model all possible orders of activities. We explain the idea using a CSP# model of the shower room scenario. All the possible activities the patient can do in the room are modelled as different choices and they are enclosed into a process named *PatientShowerRoom*.

```
PatientShowerRoom() = exitShowerRoom → PatientOutside()
          □ turnOnTap → PatientShowerRoom()
          □ turnOffTap →  PatientShowerRoom()
          □ wandering →  PatientShowerRoom()
          □ useSoap → PatientShowerRoom();
```

Here, the operator □ represents the non-deterministic choice. It operates this way that the process *PatientShowerRoom* randomly chooses an activity such as *turnOnTap* to execute. Then it may transfer control to itself again and choose *useSoap* to execute. It is guaranteed that all possible orders of activities are generated using state space exploration techniques like model checking.

However, there might exist some unrealistic orders of events. For example, there is a sequence which contains two consecutive events of *turnOnTap*. Obviously, the patient cannot perform turning tap on activity again if the tap is turned on already. In order to eliminate such cases, we need to model these constraints such that the patient's behaviour is synchronised with the status of the object being used. In fact, it is essentially the problem of modelling synchronous behaviours. We propose to use event synchronisation in CSP# and give an example of shower tap model in the following. Other ways of modelling such as using a global variable or synchronous channels are also possible.

```
ShowerTap() = turnOnTap  →  turnOffTap  →  ShowerTap();
Env() = PatientShowerRoom() ‖ ShowerTap();
```

The constraint of using tap behaviours is modelled as if *turnOnTap* event happens, it will be disabled until the *turnOffTap* activity is performed. The two processes *PatientShowerRoom* and *ShowerTap* are composed to be a complete model of the environment, *Env*. Here, the operator ‖ denotes parallel composition. Its operational semantic says that the executions of the composed processes must be synchronised on common events appearing in all of them. Interested readers can refer to [6] for more details. Here, the *turnOnTap* event becomes a common event between the two processes.

Modelling Location Transitions. While modelling the patients behaviours, we divide the activities according to the locations where they can be performed. In the *PatientShowerRoom* model, if the event *exitShowerRoom* is engaged, the process will pass control to the *PatientOutside* process. Thus, only activities outside can be selected to run while activities in the shower room are disabled. This modelling approach is to reflect the location transitions in the model and to generate realistic sequences of activities.

Modelling Multiple Users. In multiple-user sharing environment, the activities that different users can perform in a certain location are usually the same. However, in some cases, these activities need to be differentiated. For example, in AMUPADH, the system tracks different patients using RFID tags. Thus, the sitting on bed behaviour performed by patient1 and patient2 are different from the system's point of view. We model this requirement using the process parameters and events with indexes. In the following, we provide the behaviour model of the patient using bed where identify information is important.

```
PatientBed(i) = sitOnBed.i  →  PatientBed(i)
           □ lieOnBed.i  →  PatientBed(i)
           □ leaveBed.i  →  PatientBed(i);
```

Parameter i in process *PatientBed*(i) represents the identity of the patients. This identity variable is also attached to events so as to differentiate the activities performed by different patients.

3.2   Modelling System Design

PvC systems share the features such as layered architecture and concurrent communications. In the following, we discuss these common features and their modelling layer by layer.

Modelling Sensor Layer. There are a lot of interesting problems in this layer. First of all, there are different communication patterns like synchronous communication or asynchronous message passing. These communications form the basic functionality of sensors. Additionally, different sensors have different frequencies of sending messages. For example, RFID reader sends a signal to system every 1 s while pressure sensor sends every 10 s. This issue may cause the system to make wrong adaptations since the information of the environment may not be completely refreshed at some time point. Finally, sensors have limited power supply and may fail from time to time. These two problems regarding the different sending rates and unstable working conditions of sensors create many uncertainties in PvC systems.

Nonetheless, problems might also exist in the wireless network such as message loss. We skip this part since research of model checking wireless networks has been done extensively in the literature [18]. The details about signal encoding/decoding and message transmission via wireless networks are abstracted away for simplicity in our work.

Modelling Concurrent Interactions. Sensors interact with the environment by detecting events and report sensed contexts by transmitting signals to middleware. The behaviours of detecting and transmitting can be abstracted to two modelling patterns which are synchronous events and message passings respectively. Event synchronisation has been introduced in Sect. 3.1. As for message passing, there are different modelling patterns in different languages. Some languages support synchronous channels through which the sending and receiving events are synchronised. In other languages, broadcast channels or asynchronous channels with buffers are supported. In the following, we model the shake sensor using a synchronous channel.

```
channel port 0;
Shake_Sensor() = ( turnOnTap  → port!Shake.UnStationary  → Skip
       □ turnOffTap  → port!Shake.Stationary  → Skip
       ); Shake_Sensor();
```

Here, *port* is the synchronous channel defined for the shake sensor to communicate with middleware. *Shake*, *UnStationary* and *Stationary* are integer constants representing the sensor's ID and possible statuses. In the model, the shake sensor sends out the signal *UnStationary* when the tap is turned on. Note that CSP# supports multi-process synchronisation that the event *turnOnTap* can be synchronised in all three processes.

Modelling Frequency. Sensors are tuned to have different sending rates due to their functionalities and the purpose of saving energy. However, if the rates are not carefully calculated, the system may work incorrectly. To analyse these behaviours, we propose to use timed modelling languages such as Stateful Timed CSP (STCSP) [19]. Timed Automata (TA) [20] is not suitable in this case because the hierarchal modeling is not supported in TA. The modelling pattern of sending rates using STCSP would be as follows.

```
FSR_Sensor() = (sitOnBed   →⟩ port!FSR.Sitting →⟩ Skip
      ☐ lieOnBed →⟩ port!FSR.Lying →⟩ Skip
      ☐ leaveBed →⟩ port!FSR.Empty →⟩ Skip
      ☐ nothing →⟩ port!FSR.Empty →⟩ Skip
      ); Wait[10]; FSR_Sensor();
```

Here, operator →⟩ denotes the urgent event in its left hand side which cannot be interleaved by other timed events. $Wait[t]$ is the syntax to model the process idling for t time units. The above process models the periodic sensing behaviours of the pressure sensor which senses the pressure on the bed for every 10 time units. Its status is transmitted immediately after the sensing.

Modelling Sensor Failures. Sensors have limited accuracy that they may fail to detect certain events. They could also run out of battery and fail to send the signals. Intuitively, we model this with probabilistic modelling constructs, e.g., Probabilistic CSP# (PCSP#) [21], Probabilistic Timed Automata (PTA) [22].

```
RFID_Reader() = enterBedroom.1  → port!RFID.PersonA  → Skip
      ☐ enterBedroom.2  → port!RFID.PersonB  → Skip;

MalSensor() = pcase{ 9: RFID_Reader()
      1: fail  → Skip }; MalSensor();
```

Here, *pcase* is a syntax for modelling probabilities. 9 and 1 are probability weights here. This process models that the RFID reader works correctly with probability of 90 %.

In summary, different issues in the sensor layer can be modelled using different language constructs. Notice that the two modelling languages (i.e., STCSP, PCSP) we adopted are both extensions of CSP# language. As demonstrated in above examples, our intention is that it is easy to start with a simple model and extend it with richer features with minimum efforts.

Modelling Middleware Layer. As shown in Fig. 5, middleware performs the tasks of managing and reasoning contexts as well as making adaptation decisions. Messages received from sensors will trigger an update of the system knowledge/contexts. The status of a sensor is one kind of contexts. Context variables are modelled using shared variables in supporting modelling languages.

Furthermore, the reasoning engine performs reasoning by evaluating predefined rules whose conditions are propositions of context variables. A common practice for specifying rules is to use guarded processes or if-else statements. The following example models the rule in Sect. 2.3 in CSP#:

```
Rule() = if(sensors[Pressure_Sensor] == SITTING &&
    Duration[Pressure_Sensor] > 30){
    res!Act.SitTooLong.PersonA → Skip};
```

Finally, an adaptation decision will be made based on the reasoning results and sent to the application layer to execute. This again can be modelled by message passing patterns. For the above example, if the *rule* which interprets that someone is sitting on bed for more than 30 time units, a message will be sent to the application layer through the channel *res*.

Modelling Application Layer. Application layers vary according to different implementations. However, we may only care about the responsive actions which will affect the end users. Thus we focus on modelling of how the adaptation decisions are executed. For instance, in the AMUPADH system, the reminding system is modelled as follows:

```
Reminder() = res?status.rid.pid → (
            [status == Act]ActivateReminder(rid,pid)
            □[status == Deact]DeactReminder(rid,pid) ); Reminder();
ActivateReminder(rid,pid) = update{reminder[rid][pid] = true} → Skip;
```

By decoding the message received from the middleware, the workflow of reminder system diverts according to the *status* command. If it is an *Act* command, the system activates reminder *rid* to patient *pid* by calling *Activate-Reminder*(*rid*, *pid*) process. Similar logic applies for deactivating a reminder.

3.3   Compose a Complete Model

In PvC systems, different components in different layers cooperate to fulfil the system goals. However, how to model this cooperate relations are left to be discussed till now. From a careful study, we discover that in PvC systems, there are three common types of relationships between system components which are sequential, independent and concurrent relations. Sequential relation means the execution of the components is strictly sequential according to the workflows of the system. Components that are completely unrelated to each other execute independently. As for concurrently related components, they have synchronised behaviours. These relations can be well supported in hierarchical languages such as CSP#. Respectively, these three relations can be modelled as sequential, interleave and parallel compositions using operators ; , ||| and || respectively. Examples here may reuse some process names in above models. Note that parallel composition has been introduced in modelling activities in the environment.

```
Sensors() = Shake_Sensor() ||| FSR_Sensor();
Middleware() = ContextManager(); ReasoningEngine(); AdaptationManager();
```

Here, since each sensor in the environment works independently, the sensor layer model *Sensors*() is composed by the interleave operator. On the other hand, in the middleware layer, the three components are executed sequentially as determined in the workflow. Therefore, the middleware model *Middleware*() is composed using sequential operator.

Choosing a Modelling Language. The above mentioned modelling patterns are supported in most modelling languages of CSP family. It is also possible to be translated to other formalisms e.g., Timed Automata. When it comes to unify concurrent modelling with probabilistic or real time modelling or both, there exists some approaches. For example, in PAT framework, CSP# supports for modelling concurrent system behaviours; PCSP# extends CSP# with probabilistic behaviour modelling, it is suitable to model failures in PvC systems; RTCSP# extends CSP# with real time constructs which can be used to model the periodic sensing behaviour; and finally PRTS which integrates both probabilistic and real time modelling constructs under one roof. However, it is important to choose a proper modelling language according to different targets. For example, CSP# is most suitable for reasoning concurrent behaviours of PvC systems while PRTS will be an over cure. We may also argue that it requires minimal effort to extend a CSP# model to an PCSP# model and likewise.

4   Scenario Verification

After system engineers finished the design of a PvC system, they are often asked to provide guarantees for correctness and safety requirements. They may be asked to answer general questions like "Is the system free of conflict adaptations?" or "Will the services deliver when they are supposed to?". These high level requirements cannot be validated against the system thoroughly using traditional techniques like testing. However, they can be specified and verified using formal methods. For example, using model checking technique, the first question can be verified in the following steps. First, define the conflict adaption scenario as a state; secondly, using reachability verification algorithms to exhaustively search the system state space to see if such a state is reachable. In this section, we discuss the critical properties and propose their specification patterns.

4.1   Desirable Properties

Properties regarding the good behaviours of the systems are desirable.

Deadlock Freeness. Deadlock freeness is one of the important safety requirements and should be assured before checking any liveness properties. Deadlock is a situation that the system reaches a state where no more actions can be performed. It can lead to serious consequences such as falling of the patient is not being alerted to a nurse. Deadlock checking is supported in most model checkers.

Guaranteed Services. Well designed application services determine fundamental responsive behaviours of pervasive healthcare systems. For example, in a smart meeting room, upon detection of some one entered the room, a service will be scheduled to run that it will invoke an actuator to automatically turn on the lights. Effectiveness of these services is an important measurement of the system for the sake of users. To specify this requirement, we propose patterns of liveness properties using Linear Temporal Logic (LTL). For example,

$$\Box(\texttt{PatientWandering} \rightarrow \Diamond \texttt{LeaveRoomReminder})$$

Here, $\Box$ and $\Diamond$ are operators in LTL which read "always" and "eventually". This formula specifies the property meaning "Always when *Patinet Wandering* situation happens, the service *LeaveRoomReminder* will be eventually delivered".

The services are usually required to be delivered in bounded time. Obviously, it is certainly undesirable if the reminder is sent too late that even the patient has left the room. To specify the bounded liveness properties, one can use Timed Computational Tree Logic (TCTL) which extends CTL with clock constraints. The other possible solution is to bound the target system model with *deadline* semantics in some real time modelling languages such as STCSP.

Security. Since PvC systems carry lots of environment information including the user's confidential profiles, it is critical to protect privacy. Leakage of information can compromise the safety of the user and his or her belongings. For instance, food delivery person should not have access to the patients medical profile. Properties to describe security problem can be specified in many kinds of logics such as LTL. For example,

$$\Box(\texttt{FoodDeliveryPerson} \rightarrow \texttt{not} \ (\Diamond \ \texttt{AccessPatientProfile}))$$

Model checking techniques for security problems are proposed in papers such as [23].

4.2   Testing Purposes

To test the system after being deployed is cumbersome considering the reengineering workload. Fortunately, those unwanted scenarios can be specified in properties and checked using reachability verification algorithms.

System Inconsistency. Failures of sensors and wireless networks may cause contexts of the environment in the system to be out of date. Thus system knowledge can be inconsistent with actual environments. By defining such conflicting states, you can test again the system model to see if such a state is reachable.

Conflicting/False Services. To guarantee the services being eventually delivered is not enough. It is also important to check if these services are sent properly. Some problems have been reported by domain experts such as conflicts of reminders [24]. These problems are especially common in multi-user systems. For example, in AMUPADH, two conflicting reminders are prompted at the same time that one asks the patient to leave shower room while the other asks the patient to use soap to continue showering. This causes the confusion of the patient and could agitate them. Another scenario is that the reminder is sent to the wrong person. These problems can be specified in reachability properties.

5   Rules Verification

PvC systems are widely applied in healthcare domain, especially in the area of assistive living. In fact, it is challenging to automatically recognise activities of assisted people and to render adequate assistive services. In the current literature, rule based system design is adopted that it is able to provide dependable assistance services based on sensors integrated into the living ambient environment [25,26]. In such systems, rules are manually defined by system engineers based on observations from doctors and caregivers. As introduced in Sect. 2.3, a rule consists of a name, a condition field and an action field. In the condition, a certain activity to be monitored is defined based on contexts such as status of sensors, duration of sensor readings or system flag variables. The assistive service is an adaptation decision which is defined in the action field. During the reasoning process, all the rules will be evaluated based on the current contexts and actions will be executed upon the satisfaction of the rule's condition. Working in such a fashion, rule based system is able to intelligently recognise activities of users and adapt to their needs accordingly.

However, the correctness of the rules remains a non-trivial problem. Anomalies such as duplication, unreachable condition and conflicts widely exist in rule bases. Due to rule engineer may have limited knowledge of assisted user, incorrect or vague rules may be defined which will impair the system's capability in determining activities. The accuracy of activity information is lowered that it may further result in a lack of service to be offered. What's more, unreliable rules would also provide a misleading reflection of the actual situation, which is unacceptable in mission-critical or urgent scenarios. Besides, to verify relatively large rule repositories is considerably laborious. Therefore there is a need to construct an approach that is able to verify and ensure the specificity of rules, and to also provide evidence of the erroneous rules.

In this work, we propose the definition and specification of rule anomalies according to their behaviours and influence on the system behaviour (instead of using common definitions based on the syntax and semantics). By reusing the system model constructed in Sect. 3, we are able to detect rules anomalies feasibly using existing model checking algorithms. In the following, we list the three types of rule anomalies.

5.1   Non-reachable Rules

Non-reachable rules are trivial as some rules' conditions are never satisfied during all system runs. These rules can be unintentionally introduced by rule developers. Although the system's correctness is unaffected, they add complexity to the model and slow down the rules evaluation process. On the other hand, it could be the reason that a critical context used in the rule condition is always not available. This, in fact, means the system fails to detect certain events in the environment. For example, in the scenario of detecting the usage of cupboard, the rule defined for this behaviour is never fired because the sensor engineers forget to deploy a reed switch sensor (which is used for detecting open/close action). In such a case, detecting these rules will reveal critical problems of the system.

Detection of non-reachable rules can be done by reachability checking. Based on the system model, we analyse all the system states for each rule individually to see whether its condition can be satisfied or not. The pattern proposed for expressing this property is as follows.

```
assert rule_SBTLAA.condition reachable
```

We take the rule in Sect. 2.3 as an example. The rule name is represented as rule_SBTLAA (where SBTL_AA stands for sitting on bed too long for person A on Bed A). By defining its condition as a state, we try to assert whether this specific state is reachable or not. During the verification process, each system state will be compared to see if there is a match by using existing reachability checking algorithms. Non-reachable rules are better to be eliminated before checking other rule anomalies.

5.2   Redundant Rules

Redundant rules are occurrences of multiple rules firing together at same system states and producing non-conflicting system results. In fact, the rule system is usually maintained by multiple engineers, even end-users. It is often the case that they put similar rules into the system such as similar condition with different parameters or different actions. Redundant rules will increase the complexity of the rules and slow down the rule execution process. Furthermore, redundant rules create redundant information that blows up memory easily.

We define two kinds of redundant rules, *duplicated rules* and *subsuming rules*. The former refer to rules that always fire together at the same time and have a non-conflicting actions. Identical rules where their conditions and actions are both the same is considered as one special case of duplication. The latter applies to the case that one rule is always fired with the other rule where their actions are not contradict. Thus, the scenario covered by the first rule is included in the second one.

Redundant rules can be specified as LTL formulae. By the above definition, we propose the specification patterns in LTL as follows:

$$\Box(\texttt{rule1.condition} \rightarrow \texttt{rule2.condition}) \qquad\qquad (1)$$
$$\Box(\texttt{rule2.condition} \rightarrow \texttt{rule1.condition}) \qquad\qquad (2)$$

In the example, if (1) and (2) both turns out to be true, then we say rule1 and rule2 are duplicates. If only one of them is true, for instance, (1) is true, then rule1 is subsumed by rule2 where the scenarios at which rule1 satisfies is covered by rule2 as well.

5.3 Conflicting Rules

Conflict anomalies focus on the actions of rules. In a particular state, two rules both fires but with contradicting actions triggered. Then, they are considered as conflicting rules. This type of anomalies is usually not because of careless human errors, but because of limitations in the rule design. In fact, engineers define rules based on their limited knowledge of the actual user behaviour. However, it is impossible for them to figure out all the scenario. Especially in the case of dementia patient caring, the abnormal behaviour is beyond the imagination of normal people. Thus, contradictions often happen in the system. Furthermore, conflict rules are critical but difficult to detect. They could cause the user to be confused which may be harmful to their health or even life. But conflicting situations are not easily revealed during lab testing where only selected scenarios are tested. Thus, using advanced techniques such as model checking which can simulate every possible behaviour of user and perform complete search of state space are favourable.

In an attempt to reuse existing techniques, we detect conflicting rules in two ways: (1) we perform verification of Sect. 5.2. Based on the verification result, we inspect the actions of the redundant rules to find conflicting cases; (2) by defining impermissible sets which contains contradict knowledge, we check if such occasion can be reached during the system run. Note that, since the contexts only kept in the knowledge base of rule systems, contradictory contexts must imply contradicts in the rules. By performing reachability checking, we are able to find conflicts with witness traces revealing the rules which lead to the conflicting state. We take an example in AMUPADH system.

```
impermissible_set (Loc_PersonA = ShowerRoom, Status_PersonA = Sleeping)
assert impermissible_set reachable
```

In the example, we define an impermissible set says person in the shower room and person sleeping cannot be true at the same time. These two contexts both are high level contexts that are generated by rules. Thus, if two elements in the impermissible set becomes true at the same time in the model, the two rules which generate them must be conflicting in a particular scenario.

In [27], a rule modelling approach based on language translation is constructed to automate the process of rules verification. The correction strategies for rule anomalies are also proposed.

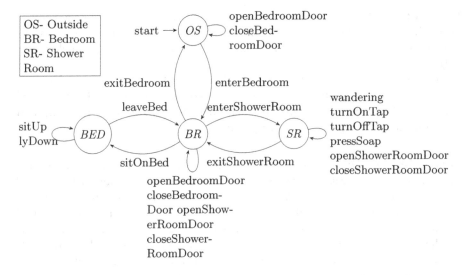

Fig. 6. Patient behaviours

6 Case Study: Formal Analysis of AMUPADH

The proposed approach is applied to analyse AMUPADH. We adopt CSP# modelling language since it supports most of the modelling patterns in the framework. Important properties are specified in reachability semantic and LTL formulae. PAT model checker is chosen to parse the model, build up the system state space and verify these properties. Experiment results are listed and unexpected bugs are reported.

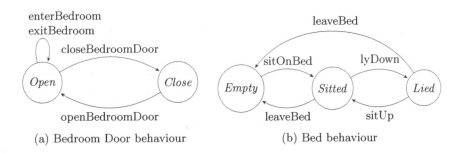

(a) Bedroom Door behaviour   (b) Bed behaviour

Fig. 7. Surrounding environment

6.1 System Modelling

In this section, we model the environments and the system design using our framework and use Labeled Transition Systems (LTS) for demonstration.

Environment Model. As shown in Figs. 6 and 7. These LTSs can be generated using simulation function of PAT. In Fig. 6, there are four possible locations that a patient can reside. The transition edges between states are labeled with patient's activities.

This patient model should be synchronised with objects within the surrounding environment. The objects that are modelled include doors of bedroom and washroom, beds and washroom taps. The behaviour models of the doors and beds are shown in Figs. 7a and b respectively.

Sensor Model. Different sensors are used in AMUPADH to monitor specific behaviours of the patients. For example, pressure sensors attached to the bed mattresses are for monitoring how the patients use the beds. The information captured by sensors is passed from sensors to the controller via a synchronised channel *port*. Every sensor possesses multiple unique states when made available to the system. Figure 8 shows the modelling of sensors using the bed RFID readers and bed pressure sensors as mentioned in Sect. 2.2. Then, we combine all processes of sensors to one process *Sensors* using composition patterns.

```
Sensors()=Rfid_Bedroom() ||| (Rfid_Beds() || FSR_Sensors())
     ||| (Rfid_ShowerRoom() || PIR_ShowerRoom()) ||| ShakeSensors();
```

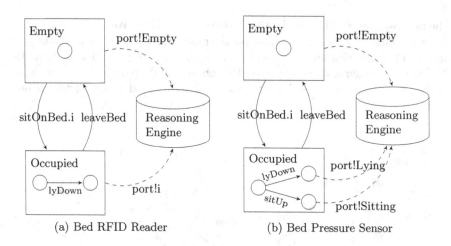

(a) Bed RFID Reader   (b) Bed Pressure Sensor

Fig. 8. Sensor behaviours

Controller and Reasoning Engine Model. Inside the reasoning engine, rule evaluation is triggered by two processes, namely the *MainInterface* and *ContextChecker* processes. In order to model the periodical evaluation by process *ContextChecker*, we use a constant integer *RATE* to represent the interval and *Duration* variable to record elapsed time. The *atomic* syntax used here is to ensure the process inside the block is executed without interference from other processes.

```
ReasonEngine()   = MainInterface() ||| ContextChecker();
MainInterface() =
        atomic{port?id.status → update{sensors[id]=status;
        Duration= call(setTimer,id,status,Duration)} →
        FireAllRules()};MainInterface();
ContextChecker()=
        atomic{update{Duration = call(tick,Duration,RATE)}
        → FireAllRules()};ContextChecker();
```

On receiving a message from any sensor, the *MainInterface* updates the sensor status and *Duration*. Then, the *FireAllRules* process is invoked to perform rules evaluation. The syntax *call(setTimer, id, status, Duration)* in the above model is used to call an external static function *setTimer* (written in C#). *Duration* be will updated externally according to the input of sensor *id* and *status*. This is a special feature in PAT, which allows users to separate complicated calculation from the high level model in order to have a simple model with efficient verification. The *ContextChecker* is similar to the *MainInterface* in updating sensor statuses and *Duration*, but does so in a periodic cycle instead of using a listener.

The process *FireAllRules* sequentially evaluates every rule independent of the results from previous cycles of rule evaluation and triggers proper actions such as setting a flag or sending a message to the reminding system. Messages are passed via a synchronous channel named *res*. We model every rule in a separate process. In the following, we list one rule to illustrate the modelling. The process *Rule_14_1()* models a complicated rule defined for recognising the wandering behaviour of the dementia patient. It says if the shake sensor on shower tap is stationary, the PIR sensor detects the patient's presence has lasted for 15 time units, the shower flag is still false and patient 1 is in the shower room, then patient1 is wandering in the shower room. Consequently, the reasoning engine sets the wander flag to true and passes a message to inform the reminding system that patient1 needs to be reminded to leave the room.

```
FireAllRules() = Rule0();
...
Rule_14_1() = if(sensors[ShakeTap] == STATIONARY &&
        sensors[PirShowerroom] == FIRING &&
        Duration[PirShowerroom] ≥ 15 &&
        !ShowerFlag && Location_Person[1] == SHOWERROOM){
        setFlag{WanderFlag = true} →
        res!Error.WanderingInShowerroom.1 → Rule_14_2()}
    else {Rule_14_2()};
...
```

Reminding System Model. In the system, reminders are activated/deactivated upon receiving corresponding messages from the controller. As shown in Fig. 9, the reminding system receives a triplet from the controller via channel *res*. This triplet consists of a command, behaviour code and patient ID. If the command is

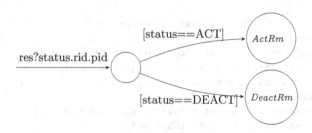

Fig. 9. Reminding system behaviours

ACT, the reminder rid will be activated and prompted to patient pid, otherwise the specified reminder will be stopped if it is active. The ACT and $DEACT$ are command constants corresponding to *Normal* and *Error* in rule processes.

Finally we integrate all the sub-system models together into a process named $SmartRoom()$ using composition patterns. Interested readers are referred to [17].

6.2   Scenario Verification Experiments

In this section we verify the proposed properties against the system model built in Sect. 6.1. He experiments test bed is a PC with Intel Xeon CPU at 2.13 GHz and 32 GB RAM.

Deadlock Freeness. Since each layer of the system as well as the environment model are independent from each other except for channel communications, we conducted the experiments incrementally. During verification of a particular component model, we abstract away the details of other component models leaving only the channels for receiving messages. Doing in this way, we are able to check deadlock freeness locally for all system components and keep the composition of component models in a manageable level. In the Table 1[3], the row starts with env represents the environment model; row $env + snr$ represents the model composed by environment model and sensor model; row $env + snr + mdw$ adds middleware model into previous one; and the last row is the complete model with all components. It turns out to be that the complete model including bedroom and shower room scenario is too large for verification. We split it into two sub-models according to the locations. The experiment results show the rapid increase of state space when more components are composed.

In CSP#, a deadlock freeness property is specified as

```
#assert CompleteSmartRoom() deadlockfree;
```

---

[3] St- States, OOM- Out of Memory.

Table 1. Results of deadlock freeness checking

Model	Bedroom		ShowerRoom		BothRooms	
	#St/k	Time/s	#St/k	Time/s	#St/k	Time/s
env	0.028	0.005	0.008	0.005	0.082	0.010
env + snr	0.157	0.080	0.072	0.030	0.906	0.339
env + snr + mdw	17.40	7.799	56.01	23.00	8319	4017
Complete	731.5	384.2	7059	4031	OOM	OOM

Guaranteed Reminders. Guaranteed reminders are important measurements of the system, which are illustrated using the patterns of service effectiveness. We take the reminder services of *Lying_Wrong_Bed* in bedroom as examples. Other properties for guaranteed reminder services can be specified similarly.

```
#define LyingWrongBed (sensors[RfidBed_1] ≠ EMPTY
         && sensors[RfidBed_1] ≠ 1);
#define RemindedWrongBed
         (ReminderStage[LyingWrongbed*2 + 1] ≠ 0);
#assert SmartBedroom() ⊨
         □ (LyingWrongBed → ◇ RemindedWrongBed);
```

Here, condition *LyingWrongBed* specifies the scenario that someone else is sleeping on patient1's bed, and *RemindedWrongBed* defines the state the reminder is prompted. This property states that when a patient is sleeping in a wrong bed, the system will always prompt the *LyingWrongBed* reminder eventually. The results of the verification are shown in Table 2. The first two reminders are checked against the bedroom system model while the rest are against shower room model. Surprisingly, all the reminders on shower room fails and it takes variant time to invalid a property due to the depth of the bugs.

Table 2. Results of guaranteed reminders checking

Property	Result	# States/k	Time/s
LyingWrongBed (LWB)	True	808.4	616.8
SitBedTooLong (SBTL)	True	798.3	607.2
ShowerNoSoap (SNS)	False	196.6	107.5
ShowerTooLong (STL)	False	1018	2635
ShowerNotOff (SNO)	False	701.8	489.1
WanderingInSR (WIS)	False	58.24	27.48

Testing of Faults. Various fault occur in AMUPADH system, the most common ones are the inconsistencies, the false reminder and reminder conflicts. They are introduced in the following. Experiment results shown in Table 3 reveals multiple bugs.

Table 3. Results of testing faults

Model	Fault type	Result	# States/k	Time/s
Bedroom	FalseAlarm: LWB	False	731.5	371.7
	FalseAlarm: SBTL	True	1.463	0.479
	CR: LWB vs. SBTL	True	20.6	7.89
Shower Room	InConsistency	True	0.404	0.180
	CR: SNS vs. WIS	True	10.34	4.150
	CR: SNS vs. STL	True	20.98	7.898
	CR: SNS vs. WNO	True	10.54	3.660
	CR: STL vs. SNO	True	16.35	5.785
	CR: STL vs. WIS	True	16.35	5.767
	CR: WIS vs. WNO	True	5.2	1.758

Inconsistent Knowledge. In the shower room, it is the case that there is no one in the room (the PIR sensor indicates *SILENT* status), while the system variable recording patient 1's location remains to be in the room. This property is specified as follows:

```
#define Contradiction ( Pos_Person[1] == SHOWERROOM
        && sensors[PIR] == SILENT);
#assert SmartShowerRoom() reaches Contradiction;
```

False Reminders. False reminders are generated prompts that should not be sent to patients. In the following, we specify a situation that the *Sit_Bed_Too_Long* reminder is sent to patient1 but in fact he is not in the bedroom.

```
#define FalseReminder (Pos_Person[1] ≠ BEDROOM
        && ReminderStage[SitBedLong] ≠ 0 );
#assert SmartBedRoom() reaches FalseReminder;
```

Conflicting Reminders (CR). In the following, *ConflictReminder* defines a state where two reminders (i.e. *WanderingInSR* reminder and *Shower_No_Soap* reminder) are simultaneously prompted to one patient.

```
#define ConflictReminder
        ( ReminderStage[ShowerNoSoap * 2] ≠ 0
        && ReminderStage[WanderingInSR * 2] ≠ 0);
#assert SmartShowerRoom reaches ConflictReminder;
```

6.3 Detecting Rule Anomalies in AMUPADH: Experiments

We perform rules verification in the following steps.

Parsing Drools Rule to CSP#. Manually modelling of all the rules are time consuming and error prone due a large number of rules are defined. Thus, we developed tool for automatically translate Drools rules used in AMUPADH to CSP# syntax in two steps.

Step 1: Extract Shared Information. For the purpose of easy management, the shared information is declared and kept in separate files from rule files. We need to first extract these information. Fortunately, customised data type and external function calls are supported in PAT. Thus, it is only needed to extend the original Java classes with additional methods for value retuning conform to PAT models. However, rewrite the Java methods to C# codes is a better solution since PAT is written in C#.

Step 2: Mapping Rules into CSP#. The parser processes the rules one at a time and splits the rule into three parts, i.e., rule name, conditions and consequences by reading the keywords *rule*, *when* and *then* respectively. We then map the Drools rule into CSP# by mapping rule names into rule associated comments, the conditions to *ifa* expressions, the consequences into *ifa* statements and point to the next rule in the *else* part (evaluation of the rules is sequential).

The two-step parsing tool automates the process of modelling rules and reduces the time required for verifying rules. Although human intervention might be required in rewriting Java classes into C#, the effort is minimal since Java classes are seldomly changed during development. However, this parser is unable to treat rules with priorities and rule chaining at the moment, extension is possible once needed.

Detetcting Rule Anomalies

Step 1: Define Rule Conditions as States. In order to specify the properties associated with evaluating conditions of rules, we explicitly define all the rule conditions as specific states. We take *Rule_14_1* in Sect. 6.1 as an example. *Rule_14_1* is the 14th rule in the rule base defined for person 1.

```
#define Rule_14_1 (sensors[ShakeTap] == STATIONARY &&
    sensors[PirShowerroom] == FIRING &&
    Duration[PirShowerroom] ≥ 15 &&
    !ShowerFlag && Location_Person[1] == SHOWERROOM)
```

Step 2: Specify Rule Anomalies. We check three types of rule anomalies.
- **Non-reachable rules** are rules that cannot be satisfied during all executions of the system. The following property asserts if *Rule_14_1* is reachable.

```
#asset SmartShowerRoom reaches Rule_14_1;
```

- **Redundant Rules** are rules that have similar effects on the system such as always fires together. We first check the subsumed rules which is the case that one rule always fires with the other rule. An simple example property specification is as follows.

```
#asset SmartShowerRoom ⊨ □ (Rule_1 → Rule_2);
#asset SmartShowerRoom ⊨ □ (Rule_2 → Rule_1);
```

By checking every pair of rules, we are able to do a thorough testing to find all possible subsumed rules. Duplicated rules are two rules subsumed with each other.

- **Conflicting Rules** are rules that fires together at a particular state but have conflict consequences. To detect this anomaly is to define an impermissible set like the following example.

```
#define contradict_state (ShowerFlag == true &&
    Location_Person[1] != SHOWERROOM &&
    Location_Person[2] != SHOWERROOM);
#assert SmartShowerRoom reaches contradict_state;
```

This *contradict_state* says someone is taking shower, but neither of the two patients is in the shower room. Note that there are only two users in the model and *ShowerFlag*, *Location_Person*[1] and *Location_Person*[2] are all variables updated by some rules. If this contradict state is reachable, there must be some rules conflicting. By inspecting the witness trace reported by PAT, we are able to identify them.

Step 3: Verify Rules Using PAT. Finally, we integrate all the rules into the formal model and verify the properties specified in previous step using built-in verification algorithms in PAT. The results are shown in Table 4 with multiple conflicts found.

Table 4. Results of detecting rule anomalies

Scenario	# Rules	# Non-reachable	# Subsumed/# Duplicated	# Conflict
Bedroom	17	2	8/3	2
Shower Room	22	5	16/2	4
Avg. Time(s)	-	2.05	3.05	-

6.4   Bug Report

Discovery of Unexpected Bugs. Counterexamples are returned as evidences if the system model violates certain properties. They are of great value to system engineers to debug the system. The set of confirmed bugs are reported as follows which are unexpected by the development team.

System implementation fails to meet requirements

- *Guaranteed Reminders.* This experiment reveals a critical problem of the system that the system fails to monitor the patient's location correctly. A patient exiting the shower room with tap left on is a typical case. The two reminders, *Shower_Not_Off* and *WanderingInSR* will repeatedly prompt even though there is no one in the shower room.

Unexpected Faults Arising out of system complexity

- *False Alarm in Bedroom.* The result of the second property is witnessed to be valid. Through careful investigation, we notice that the rule defined for *Sit_Bed_Too_Long* does not have an identity attached to the rule's condition and hence this reminder is sent to the bed's default owner regardless of the bed's current user.
- *Conflict Reminders.* From the experiment results, we found many scenarios where there are reminder conflicts. For example, a patient wandering in the shower room tirggers the *WanderingInSR* reminder. He then ignores the reminder and turns on the shower tap to play with water (A typical behaviour of a dementia patient). The water runs for a long time that the *Shower_No_Soap* reminder is triggered, therefore causing the system to prompt the conflicting reminders.

Anomalies in Activity Recognition Rules

- *Non-reachable rule.* In the bedroom scenario, the rule defined to recognise an activity of opening a cupboard is not reachable. The reason is that sensor engineers removed the reed switch sensor on the cupboard without notifying the rule engineers.
- *Redundant rules.* Five duplicated rules are discovered which were accidentally added into the rule repository for testing and were not removed due to negligence.
- *Conflicting rules.* A scenario where a pair of conflicting rules are witnessed that the monitored user have been showering for a long period of time, yet continues to ignore the reminder that prompts him to use the shower foam. This was the reason that led to the triggering of two contradictory reminders that request a user to perform activities in two different locations at the same time, which is physically impossible.

Discussion

Usefulness. We gained several observations from this case study. First, model checking techniques can provide a very good guide on system design. From our experiences of working with designers of the system, they usually focus on setting up a demonstration based on selected scenarios without considering other useful situations. It is not only because of the high cost of hardware devices but also to complete a full demonstration is time consuming. In fact, the development and consideration of all possibilities when constructing scenarios and rules is an impossible task and would either take many man-hours to find out through actual deployment. In fact, some of the bugs (e.g., False Alarm) we reported are occurring in execution of AMUPADH system and some of them are unexpected (e.g., inconsistency). The counterexamples reported from the experiment also helped the engineers to pinpoint the source of the bug. Besides, it is important to find unexpected bugs based on the stakeholders requirements before deployment of the whole system. Hence the engineers can retrieve certain normal or abnormal scenarios they are interested in based on our analysis results.

Additionally, we observed the failure of updating the correct location information of the patient leads to the violation of important properties. From the discussion with hardware engineers, we learned that RFID readers have limited detection range. We may think it is unwise to solely rely on RFID readers to track the patients. During the experiments, we also noticed that a lot of redundant messages are sent out by the reasoning engine which increase the complexity of the system and slow down the verification.

Thoughts of solving state space explosion in PvC system verification. The experimental results reflect the typical state space explosion problem. The number of states in checking deadlock freeness of the complete model reaches the level of 10^8, which is the limit of explicit-state model checkers like SPIN and PAT. The state of art state space reduction methods like partial order reduction may not have significant improvement of this problem. Compositional verification on the other hand draws our attention. From the deadlock freeness checking, we noticed that if all components are locally deadlock-free, it is of great possibility that the complete system model which is a composition of all the components is free of deadlock. Obviously verifying a local property of a component is much easier than verify it against the system model. Furthermore, the general architecture of PvC systems suggests that there are almost no sharing recourses between components. The independency between system components further proves that compositional verification could be a feasible solution to state space explosion problem. Thus, in future, we shall explore how composition verification techniques can be applied.

7   Related Work

PvC systems have achieved many milestones in recent years. However, works on applying formal methods to assure the correctness of such systems are limited. In [7], they proposed a TCOZ model for a smart meeting room system which very well captured the synchronised communications and real-time constraints of sensors and actuators. Researchers in [8] used Ambient Calculus to model a location sensitive smart guiding system in a hospital. The mobility issue is well modelled in their work. Important properties are manually proved in both of the two papers. However, both of the two languages does not have support for hierarchical structures. Moreover, lack of verification tools support restricts the applicability of their approaches to large pervasive systems. Our work advances them by adopting hierarchical modelling patterns. Automatic verification of our modelling framework can be supported by popular model checkers.

In [28], Adaptation Finite-State-Machine (A-FSM) is proposed for modelling adaptations between system states in context aware adaptive applications. Fault patterns based on the A-FSM and their detection algorithms are presented as well. However, how to model systems in A-FSM is not clear and liveness properties are not supported in their work. Researchers in [11] proposed multiple important properties regarding security, safety requirements in PvC systems. Formalisation patterns are illustrated and possible verification approaches are explored. However, since they lack the underlying modelling patterns, the properties and their verifications are very difficult to apply. In our work, we further classified the important requirements into safety and liveness properties and formalise them in popular logics which are checkable based on our modelling framework. Besides, we propose scenario verification which verifies critical requirements on an exhaustive enumeration of targeted scenarios which is more focused than aimless, random verification approaches and more complete than verification/testing upon selected cases.

In rules verification, Ligeza and Nalepa [13] proposed definitions for rule anomalies regarding redundancy, consistency, completeness and determinism. Preece et al. [14] surveyed the verification of rule based systems focusing on detecting anomalies. Five rule verification tools are compared based on their capability of detecting rule anomalies such as redundancy, ambivalence etc. However, their definitions for anomalies and the surveyed algorithms are not directly applicable to PvC systems. Most of their algorithms detect anomalies based on syntax checking and semantic logics inspection between rules, instead of how the rules affect the system behaviour. Furthermore, the algorithms are mostly designed for goal-driven (stateless) rules where knowledge is not shared between different rounds of evaluation. This is certainly not the case of how rules working in PvC systems. Thus, in our work, we redefined the rule anomalies according to their influences upon the system behaviours and formulate them into properties which are verifiable on our modelling framework by reusing existing model checking algorithms.

8   Conclusion

In this work, we propose a formal modelling framework for pervasive computing systems. Different modelling patterns are discussed according to the typical features of systems such as concurrent interactions, context-awareness and layered architectures. We also provide environment modelling patterns which are usually not considered in modelling complex systems. Based on the modelling framework, we propose scenario verification where critical properties of safety and liveness requirements are identified and specified in proper logics such as specifying guaranteed reminder services using LTL, and rules verification where rule anomalies are redefined upon system behaviours and formulated to formal properties which can be verified using existing model checking algorithms.

To demonstrate our approaches, we present a case study of an living assisting system for elder dementia patients. We model the system using our modelling framework and conduct experiments of scenario verification and rules verification. Multiple bugs are revealed. Experimental results and sources of the bugs are explained.

This work demonstrates the usefulness of formal methods (particularly model checking techniques) in analysing PvC systems. In the future, we will apply probabilistic model checking techniques for quantitative analysis of PvC systems and explore compositional verification techniques to alleviate the state space explosion problem.

Acknowledgment. The authors would like to thank Lee Vwen Yen Alwyn, Clifton Phua, Zhu Jiaqi and Kelvin Sim from Institute for Infocomm Research in Singapore for the kindness contributions and valuable feedback to this work. We also want thanks the anonymous reviews for their valuable suggestions in improving the manuscript.

References

1. Weiser, M.: The computer for the 21st century. Sci. Am. **265**(3), 66–75 (1991)
2. Estrin, D., Culler, D., Pister, K., Sukhatme, G.: Connecting the physical world with pervasive networks. IEEE Pervasive Comput. **1**(1), 59–69 (2002)
3. Nehmer, J., Becker, M., Karshmer, A., Lamm, R.: Living assistance systems: an ambient intelligence approach. In: Proceedings of the 28th International Conference on Software Engineering, ICSE '06, pp. 43–50 (2006)
4. Saha, D., Mukherjee, A.: Pervasive computing: a paradigm for the 21st century. Computer **36**, 25–31 (2003)
5. Edwards, W.K., Grinter, R.E.: At home with ubiquitous computing: seven challenges. In: Abowd, G.D., Brumitt, B., Shafer, S. (eds.) Ubicomp 2001. LNCS, vol. 2201, pp. 256–272. Springer, Heidelberg (2001)
6. Sun, J., Liu, Y., Dong, J.S., Chen, C.: Integrating specification and programs for system modeling and verification. In: TASE, pp. 127–135 (2009)
7. Dong, J.S., Feng, Y., Sun, J., Sun, J.: Context awareness systems design and reasoning. In: ISoLA, pp. 335–340 (2006)
8. Coronato, A., Pietro, G.D.: Formal specification of wireless and pervasive healthcare applications. ACM Trans. Embed. Comput. Syst. **10**, 12:1–12:18 (2010)
9. Mahony, B., Dong, J.S.: Blending object-Z and timed CSP: an introduction to TCOZ. In: ICSE '99, pp. 95–104 (1998)
10. Cardelli, L., Gordon, A.D.: Mobile ambients. In: Nivat, M. (ed.) FOSSACS 1998. LNCS, vol. 1378, pp. 140–155. Springer, Heidelberg (1998)
11. Arapinis, M., Calder, M., Denis, L., Fisher, M., Gray, P.D., Konur, S., Miller, A., Ritter, E., Ryan, M., Schewe, S., Unsworth, C., Yasmin, R.: Towards the verification of pervasive systems. ECEASST **22**, 1–15 (2009)
12. Clarke Jr, E.M., Grumberg, O., Peled, D.A.: Model Checking. MIT Press, Cambridge (1999)
13. Ligeza, A., Nalepa, G.J.: Rules verification and validation. In: Giurca, A., Gasevic, K.T.D. (eds.) Handbook of Research on Emerging Rule-Based Languages and Technologies: Open Solutions and Approaches, pp. 273–301. IGI Global, Hershey (2009)

14. Preece, A.D., Shinghal, R., Batarekh, A.: Principles and practice in verifying rule-based systems. Knowl. Eng. Rev. **7**(02), 115–141 (1992)
15. Biswas, J., Mokhtari, M., Dong, J.S., Yap, P.: Mild dementia care at home – integrating activity monitoring, user interface plasticity and scenario verification. In: Lee, Y., Bien, Z.Z., Mokhtari, M., Kim, J.T., Park, M., Kim, J., Lee, H., Khalil, I. (eds.) ICOST 2010. LNCS, vol. 6159, pp. 160–170. Springer, Heidelberg (2010)
16. Sun, J., Liu, Y., Dong, J.S., Pang, J.: PAT: towards flexible verification under fairness. In: Bouajjani, A., Maler, O. (eds.) CAV 2009. LNCS, vol. 5643, pp. 709–714. Springer, Heidelberg (2009)
17. Liu, Y., Zhang, X., Liu, Y., Sun, J., Dong, J.S., Biswas, J., Mokhtari, M.: Technical report for formal analysis pervasive computing systems. http://www.comp.nus.edu.sg/~yanliu/techreport.pdf
18. Olveczky, P.C., Thorvaldsen, S.: Formal modeling, performance estimation, and model checking of wireless sensor network algorithms in real-time maude. Theor. Comput. Sci. **410**, 254–280 (2009)
19. Sun, J., Liu, Y., Dong, J.S., Liu, Y., Shi, L., Andre, E.: Modeling and verifying hierarchical real-time systems using stateful timed CSP. ACM Trans. Software Eng. Methodol. (TOSEM) **22**(1), 3:1–3:29 (2013)
20. Alur, R.: Timed automata. Theor. Comput. Sci. **126**, 183–235 (1999)
21. Sun, J., Song, S., Liu, Y.: Model checking hierarchical probabilistic systems. In: Dong, J.S., Zhu, H. (eds.) ICFEM 2010. LNCS, vol. 6447, pp. 388–403. Springer, Heidelberg (2010)
22. Kwiatkowska, M., Norman, G., Parker, D.: PRISM 4.0: verification of probabilistic real-time systems. In: Gopalakrishnan, G., Qadeer, S. (eds.) CAV 2011. LNCS, vol. 6806, pp. 585–591. Springer, Heidelberg (2011)
23. Marrero, W., Clarke, E., Jha, S.: Model checking for security protocols. Technical report. Carnegie Mellon University (1997)
24. Du, K., Zhang, D., Zhou, X., Hariz, M.: Handling conflicts of context-aware reminding system in sensorised home. Cluster Comput. **14**, 81–89 (2011)
25. Antoniou, G.: Rule-based activity recognition in ambient intelligence. In: Bassiliades, N., Governatori, G., Paschke, A. (eds.) RuleML 2011 - Europe. LNCS, vol. 6826, pp. 1–1. Springer, Heidelberg (2011)
26. Storf, H., Becker, M., Riedl, M.: Rule-based activity recognition framework: challenges, technique and learning. In: PervasiveHealth, pp. 1–7 (2009)
27. Lee, V.Y., Liu, Y., Zhang, X., Phua, C., Sim, K., Zhu, J., Biswas, J., Dong, J.S., Mokhtari, M.: ACARP: auto correct activity recognition rules using process analysis toolkit (PAT). In: Donnelly, M., Paggetti, C., Nugent, C., Mokhtari, M. (eds.) ICOST 2012. LNCS, vol. 7251, pp. 182–189. Springer, Heidelberg (2012)
28. Sama, M., Elbaum, S., Raimondi, F., Rosenblum, D.S., Wang, Z.: Context-aware adaptive applications: fault patterns and their automated identification. IEEE Trans. Softw. Eng. **36**, 644–661 (2010)

Revisiting Agent-Based Models of Algorithmic Trading Strategies

Natalia Ponomareva(✉) and Anisoara Calinescu

Department of Computer Science, University of Oxford, Oxford, UK
natalia_ponomareva@yahoo.com, Ani.Calinescu@cs.ox.ac.uk

Abstract. Algorithmic trading (AT) strategies aim at executing large orders discretely, in order to minimize the order's impact, whilst also hiding the traders' intentions. The contribution of this paper is twofold. First we presented a method for identifying the most suitable market simulation type, based on the specific market model to be investigated. Then we proposed an extended model of the Bayesian execution strategy. We implemented and assessed this model using our tool AlTraSimBa (ALgorithmic TRAding SIMulation BAcktesting) against the standard Bayesian execution strategy and naïve execution strategies, for momentum, random and noise markets, as well as against historical data. Our results suggest that: (i) *momentum market* is the most suitable model for testing AT strategies, since it quickly fills the Limit Order book and produces results comparable to those of a liquid stock; (ii) the priors estimation method proposed in this paper − within the Bayesian adaptive agent model − can be advantageous in relatively stable markets, when trading patterns in consecutive days are strongly correlated, and (iii) there exists a trade-off between the frequency of decision making and more complex decision criteria, on one side, and the negative outcome of lost trading on the agents' side due to them not participating actively in the market for some of the execution steps.

Keywords: Algorithmic trading · Bayesian adaptive agents · Simulation · Backtesting

1 Introduction

Algorithmic trading (AT) strategies, which model decisions made by the algorithms as opposed to human actors [28], have gained significant strength in the recent years [20]. According to [20], "for large stocks AT narrows spreads, reduces adverse selection, and reduces trade-related price discovery. Their findings indicate that AT improves liquidity and enhances the informativeness of quotes". Similarly, [15] found AT to be a cost-effective technique for large orders.

Market activity is comprised of the execution of orders submitted by the market participants, and the most common types of orders are *market orders* and *limit orders* [31]. A *market order* is the order to be executed immediately,

© Springer-Verlag Berlin Heidelberg 2014
R. Kowalczyk and N.T. Nguyen (Eds.): TCCI XVI, LNCS 8780, pp. 92–121, 2014.
DOI: 10.1007/978-3-662-44871-7_4

at the best available price on the market. A *Limit order* is defined by additionally providing a limit (or activation) price, *i.e.*, the price which, once reached by the market, will activate the order and lead to its execution. The limit price for the *Sell* order is above the current price, and for the *Buy* order it is below the market price.

Execution trading algorithms, which we will focus on, aim to execute an order submitted by a human, such as to minimize both the associated transaction cost and the market impact of the order. The two most common approaches for evaluating trading strategies are *backtesting* [21,28] and *evaluation via artificial market simulation* [10,21,24]. In *backtesting* the algorithms are evaluated using historical market data [21,28]. For each discrete time point t, the agents have access to historical data on the current price at time t. The agents submit two different types of orders: *market orders* and *limit orders*. Market orders are executed at the current price. Limit orders are either executed at the current price, if their limit price allows immediate execution, or are saved into the *Limit Book*. Then time is incremented, $(t + 1)$, the next current price is read from historical data, and the agents repeat the previous steps. This iterative process continues until all historical data items are processed. Backtesting of AT strategies is acceptable under the assumption that the agents' orders have little or no market impact, and would not have changed the historical prices if they were executed within the real market.

In *agent-based simulations* the agents representing humans and AT strategies compete with each other and determine the price of the stocks [10,21,24]. Since price fluctuations depend on the interactions of all the agents, and additional conditions that did not take place in historical data can occur, trading strategies can be arguably more fully evaluated by this approach than in the backtesting model. Bonabeau [10] state that an agent-based model that simulates the effect of regulatory changes on markets, under various market conditions and for different agent capabilities and strategies, has been used by NASDAQ to evaluate the impact of tick-size reduction. This model revealed a surprising behaviour, such as the fact that a reduction in the market tick's size can reduce the market's ability to perform price discovery, thus leading to an increase in the bid-ask spread.

The contribution of the work presented in this paper is twofold. First we investigated different types of agent-based market simulations and suggested how to identify the most suitable market simulation type, based on the specific market model to be investigated.Then we proposed an extended model of the Bayesian execution strategy. We implemented and assessed this model using our AlTraSimBa (ALgorithmic TRAding SIMulation BAcktesting) tool against the standard Bayesian execution strategy and naïve execution strategies, for momentum, random and noise markets, as well as against historical data. The results revealed valuable insights on the trade-offs between the frequency of decision making and more complex decision criteria, on one side, and the negative outcome of lost trading on the agents' side due to them not participating actively in the market for some of the execution steps.

The remainder of this paper is organised as follows. Section 2 reviews agent-based simulation models within an AT context. Section 3 describes architectural challenges to consider and summarizes some of the features of our tool. Section 4 reviews algorithmic execution agents, performance metrics, the standard adaptive Bayesian execution algorithm, and presents the theoretical extension we proposed to this approach. Section 5 presents and analyses our results and assesses our approach. The last section summarises and concludes the paper, and suggests future work directions.

2   Agent-Based Models of Financial Markets

In order to build a realistic market model and simulation, one must decide the algorithm which will model how the price of the stocks will vary over time, and the type of agents that will participate in the market. Once these modelling decisions have been made, the model implementation and architectural decisions pose several challenges, too. This section reviews the different features that are relevant for agent-based modelling and simulation, within a financial market context.

2.1   Continuous Double Auction

Price determination specifies how the prices of securities change within a market context where supply and demand do not match each other [24]. The most realistic way to simulate real markets is the Continuous Double Auction (CDA) market [24,33], where the agents are allowed to submit any type of orders (*Buy* or *Sell*), at any time.

CDA maintains an *Order Book* containing the list of the submitted limit orders. For each type of order (*Buy* and *Sell*) there is a separate queue. In the *Buy* queue the orders are sorted in decreasing order of limit price; if several limit orders have the same limit price, they are stored using a FIFO criterion. The *Sell* queue orders are sorted in increasing order of both limit price and their arrival time. The first order in the *Buy* queue determines *the bid price*, *i.e.* the highest price at which buyers are willing to purchase securities. The first order in the *Sell* queue dictates *the ask price* – the lowest price which sellers are willing to accept [23,33].

A submitted market order is executed against the *Limit Order Book* of an opposite direction. If the market order is larger than the limit order at the head of the queue, the next limit order from the queue is retrieved. Similarly, if a limit order arrives, it is first checked against a queue of opposite direction, to see if its immediate execution is possible, *i.e.*, if a buy limit order has a limit price more or equal to the current ask price, or if a sell limit order's price is less or equal to the current bid price. If a limit order cannot be executed yet, it is saved into the queue of the order's direction. Due to the nature of this mechanism, *the bid price is always lower than the ask price*.

CDA assumes order divisibility, *i.e.*, an order can be executed partially – any agent willing to buy (sell) stocks agrees to have only part of the order executed, if necessary [33].

2.2 Human-Like Agents

Human-like agents are used to model human actors on the market, to generate trading volume and liquidity, and fill the limit order book - *e.g.* initialize the simulation for testing AT strategies. These agents need to generate order flows comparable to those in real financial markets. For this purpose, four main human-like agent types are proposed in the literature: *noise agents, random trading agents, momentum trading agents* and *contrary trading agents* [21].

Noise Agents (Zero Intelligence). Many researchers choose to use zero intelligence agents in their financial market simulations, for example in the models presented in [12,13,18,28]. There is a strong empirical evidence backing up such a widespread use of noise agents - [18] has shown that noise agents with parameters derived from empirical data can explain up to 96 % of the variance of spreads and 76 % of the diffusion rates. There are different variations of zero-intelligence agents, the simplest ones ranging from submitting buy or sell and market or limit orders with equal probabilities, to extended models like those in [14].

Noise agents represent uninformed traders on the market. There exist a number of noise agent variations. Relatively complicated strategies for noise traders are used in [12–14]. In these papers, the agents can submit *Buy* and *Sell* orders with equal probabilities, and can submit *Limit* and *Market* orders, or can cancel existing an limit order if it was not yet executed. For limit orders, limit prices can lie inside the current spread, outside the current spread, or they can be equal to the current spot price. The inside-the-current-spread limit price is derived using the uniform distribution, and the outside-the-current-spread limit price is derived using the power law. The size of the market and limit orders follow power law distributions, with different parameters. The corresponding set of parameters suggested by [13] is presented in Table 1.

Equation 1 illustrates how a random value can be obtained from the power law distribution, with parameters A and α.

$$y = A/x^{\frac{1}{\alpha-1}}, \quad x \sim uniform(0,1) \tag{1}$$

Random Trading Agents. Random trading agents can submit either a *Sell* or a *Buy* order, each with probability 0.5 [21]. In contrast with noise agents, random agents never submit market orders, instead they always submit limit orders for a fixed number of shares. The random trading agent's logic is summarized in Table 2, where α_t is a random variable $\sim uniform(-0.1, 0.1)$ and p_{t-1} is the previous price of the security [21].

Table 1. Noise agent parameterization

Parameter	Value
Order cancellation probability	0.5
Market order probability	0.16
Market order size	$\sim powerlaw(10, 2.7)$
Limit order probability	0.35
Limit order in spread probability	0.32
Limit order at spot price probability	0.33
Limit order outside the spread probability	0.35
Limit price inside the spread	$\sim uniform(bid, ask)$
Limit price outside the spread	$\sim powerlaw(50, 2.5)$
Limit order size	$\sim powerlaw(10, 2.1)$

Table 2. Random trading agents parameterization [21]

Order type	Action
Buy order	Submit Limit Buy order with quantity N and limit price $(1 + \alpha_t)p_{t-1}$
Sell order	Submit Limit Sell order with quantity N and imit price $(1 + \alpha_t)p_{t-1}$

Momentum Trading Agents. Momentum trading agents are trend followers – they assume the persistence of an observed (so far) trend and act based *on this belief* [21]. For example, if the stock price went up compared to the previous observations, momentum agents buy this stock, assuming that it will continue to appreciate in price.

Contrary Trading Agents. Contrary agents have an opposite opinion and assume that a trend is just a temporary inconsistency in the market, perhaps an overreaction to a new announcement, and thus act *against* the trend [21]. For example, they choose to sell stocks if there was an upwards trend, and buy if there was a downwards price trend. The strategies of momentum and contrary trading agents are presented in detail in [21], and are summarised in Table 3.

3 Implementation

3.1 Essential Features of the AT Simulation Tool

Once the features and parameters of the model are decided, the next important question that arises is what features should the simulation tool include, and how it should be implemented. Based on the analysis and evaluation of several financial market simulation tools [12,19,21,29,32] and available agent-based platforms like SWARM [6] and JASA [4], we identified the following list of features that a suitable tool for evaluating AT strategies should have:

Table 3. Momentum and contrary agents parameterization [21]

Condition	Action
Momentum agent	
$p_{t-1} < p_t$	Submit Limit Buy order with quantity N and limit price $(1 + \alpha_t)p_{t-1}$
$p_{t-1} > p_t$	Submit Limit Sell order with quantity N and limit price $(1 + \alpha_t)p_{t-1}$
Contrary agent	
$p_{t-1} < p_t$	Submit Limit SELL order with quantity N and limit price $(1 + \alpha_t)p_{t-1}$
$p_{t-1} > p_t$	Submit Limit BUY order with quantity N and limit price $(1 + \alpha_t)p_{t-1}$

- The ability to switch between backtest mode and CDA mode, like in [28], thus rendering the AT simulation tool suitable for both testing the performance of AT strategies against historical data, as well as allowing to explore how the agents and algorithms will behave and what patterns might emerge from the agent interactions for CDA-based simulations.
- The ability to simulate real market features – while this requirement adds complexity, these features might be essential to understand what strategy will perform best in a real market context. They include:
  - *Short selling.* Short selling is essential because allowing algorithms to model short trading can substantially extend their power, and thus more complicated algorithmic agents can be evaluated.
  - *Commisions* are also a prerequisite, because modelling them can have a substantial impact on the agents' performance.
  - *Lower latency* - as AT strategies can process information, make decisions and execute orders considerably faster than humans.
- Finally, as suggested in [32], the system should be of a pluggable architecture and easily extensible, to allow the addition of new agents and new modules.

An important part of implementation is to decide how the agents will behave during simulations. Many implementations, like [19], choose an approach where at each point of time only one agent, selected randomly, will be able to perform any market actions. This is usually simulated by selecting an agent number from a distribution, and then giving this agent opportunity to act. This, however, makes it complicated to simulate the lower latency of algorithmic agents as, for this feature to be simulated, algorithmic agents should have a higher probability of being selected.

We, however, chose to adopt a different approach: at each time all agents can act simultaneously. In the implementation of our simulation tool, each agent is represented by a thread and can make decisions and submit orders at each auction time moment (*price tick*). Agents are synchronized only when they are submitting orders, because it is impossible to submit two orders simultaneously and there is always an order submitted earlier. When the next tick occurs, all agents are notified and can choose either to skip the step or act at this time. The probability of skipping the step should be higher for human-based agents than the probability of skipping the step for algorithmic agents, thus simulating the lower latency.

In the next section we briefly describe the architecture we used for our simulation tool, AlTraSimBa, in terms of its modules and how they are linked, and we explain the reasoning behind these design and implementation decisions.

3.2 AlTraSimBa Implementation

There are two separate components of AlTraSimBa: a core component and a UI component, in charge of graphical user interface. In the core part the main modules are Auction, PriceDetermination, ClearingHouse and HistoricalData-Log. Figure 1 demonstrates the schematic view of the modules' interaction.

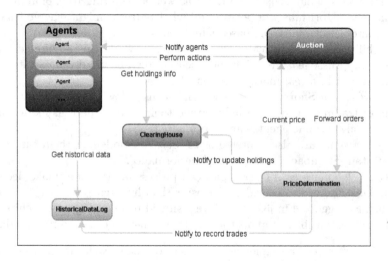

Fig. 1. Modules of AlTraSimBa system

Auction is the core module in charge of time iteration, notification of agents about auction events (tick time passed, auction day passed, etc.), collection of orders and directing them to the appropriate price determination module (either CDA or BackTest). It is implemented as a separate thread that communicates with the Agent threads via observer–like communication (all agents are listeners of auction events).

PriceDetermination is in charge of price formation and models the market mechanism. For CDA, price changes are determined by the executed orders on the market, while for BackTest prices follow prefixed fluctuations from the loaded data file. This module executes market orders, maintains the *Limit order book*, and executes limit orders when limit price conditions are met. The PriceDe-termination module is connected to the ClearingHouse module, to be able to inform the latter that the order of an agent was executed, and therefore that the agent's holdings should be updated. Additionally, this module notifies HistoricalDataLog when trading takes place and forwards the details of the trade to be recorded.

HistoricalDataLog contains the information about the trades that took place on the market in the past. It also makes historical price movements and trading volumes available to all the agents in the model, for training and decision-making purposes.

ClearingHouse is in charge of keeping agent-specific information: the current holdings of an agent (the number of stocks owned, free cash and level of credit, number of shares shortsold), initial wealth level, and the list of submitted, cancelled, executed or still outstanding orders.

Even though we tried to implement as many key features as possible, due to the limited time AlTraSimBa suffers from a number of shortcomings which should be addressed in the future work. These limitations are:

- Borrowing is interest-free in our implementation. Ideally interest should be added for each period of having negative balance, however it requires maintaining a history of holdings and applying interest accordingly. This real market feature is left for later implementations.
- Short selling does not incur commissions. By contrast, in the real world short selling requires to have a margin account with holdings that should not fall below a certain level of the price of all borrowed securities, otherwise a margin call is encountered [31]. In the future versions of the tool interest and margin calls should be implemented.
- The AlTraSimBa agents can so far submit market and limit orders only, other types of real orders are not implemented. Adding agents who can submit, for example, stop and stop limit orders can result in more sophisticated and potentially profitable strategies.

Even though the AlTraSimBa simulation tool does not incorporate all real market features, it nevertheless provides a powerful and flexible functionality for the complex evaluation of different algorithmic trading strategies, on different types of markets, and can therefore be valuable for both researchers and practitioners.

4 Execution Algorithmic Agents

In this section we explain why execution strategies are important, review metrics used to assess these strategies, and list the main relevant execution strategies. Finally, we summarize the Bayesian execution agent with daily cycle strategy and describe our theoretical extension to this strategy.

4.1 Background

An order of almost any size can be executed via the market order. However, a large market order will consume limit orders from the limit order book, step by step, from the top of the queue, obtaining worse prices as deeper limit orders are executed to satisfy the market order. Execution algorithms aim at executing large orders in such a way as to minimize the market impact of the order. They generally work by slicing large orders into smaller orders (suborders) and submitting the latter to the market at discrete times.

The problem of an optimized trade execution can be formulated as follows: the algorithm is supposed to buy or sell a specified number of shares of a specified stock, over a predetermined horizon, while taking into account the execution goals. Execution goals can be maximizing the revenue (for *Sell* orders) or minimizing the cost (for *Buy* orders), or aim to meet other investor performance metrics (*e.g.*, speedy execution, minimization of market impact). The trading strategy controls decisions about how to best split the order into smaller orders, the timing for submission of each suborder, what type of order to use, and how aggressively priced should the suborders be if limit orders are used [12].

Execution strategies can be divided into strategies that aim to meet market benchmarks and strategies aiming to meet criteria imposed by the order itself, at the moment of order submission [19].

4.2   Execution Strategy Evaluation

The objectives of an execution strategy dictate the performance metrics to be used for its evaluation. The algorithms focused on meeting market benchmarks could be assessed using the *Volume Weighted Average Price* (*VWAP*) metric [13,23].

The *Volume Weighted Average Price* (*VWAP*) is the average price on a per share basis over a specified period H, for example over one trading day [13,23]. If during the horizon H, T trades took place on the market $\{(p_1, v_1), \ldots, (p_T, v_T)\}$, where p_i denotes the price at which the trade was executed and v_i the traded volume, then the market $VWAP_M$ is defined as:

$$VWAP_M = \sum_{t=1}^{T} p_t v_t \Bigg/ \sum_{t=1}^{T} v_t \qquad (2)$$

The *VWAP* of some algorithm A is defined analogously using the sequence of orders algorithm executed. To compare the market and algorithm's metrics, the VWAP Ratio (*VWAPR*) is used [23]:

$$VWAPR = \begin{cases} \dfrac{10000 \times (VWAP_A - VWAP_M)}{VWAP_M} & \text{for a } Buy \text{ order} \\ \dfrac{10000 \times (VWAP_M - VWAP_A)}{VWAP_M} & \text{for a } Sell \text{ order} \end{cases} \qquad (3)$$

The smaller the *VWAPR*, the better is considered the strategy. A negative *VWAPR* means that the $VWAP_A$ was better than $VWAP_M$ [23]. Apart from just the *VWAPR*, [13] also considers the algorithm's *VWAPR* mean and standard deviation over N trading days. Low *VWAPR* mean and standard deviation values are expected from a good execution strategy aimed at achieving the *VWAP* metric.

Another metric, the *Mid-Spread Metric* (*MSM*) is calculated by comparing the overall execution price achieved by an execution strategy to the average of the bid-ask spread at the time when the order was submitted to the strategy for execution [26]. *MSM* provides a good baseline for the comparison of different execution algorithms that do not aim at achieving market-specific metrics. In general, any real strategy is expected to do worse than this idealized policy (*MSM* is also referred to as *underperformance* or *trading cost*) [25], unless market prices will move in a direction that favours the trader. *MSM* is measured in basis points (1/100 unit of percent); the *MSM* equation is given below:

$$MSM = \begin{cases} \dfrac{10^4 * (\sum_{t=1}^{K} p_t n_t + C(K-1) - p_{mid} * X)}{p_{mid} * X} & \text{for a } Buy \text{ order} \\ \dfrac{10^4 * (p_{mid} * X - \sum_{t=1}^{K} p_t n_t - C(K-1))}{p_{mid} * X} & \text{for a } Sell \text{ order} \end{cases} \quad (4)$$

where $\{(p_1, n_1), (p_2, n_2), \ldots, (p_K, n_K)\}$ is the sequence of suborders executed by the strategy, p_{mid} is equal to $(p_{ask} - p_{bid})/2$, $X = \sum_{t=1}^{K} n_t$ is the volume of original (large) order, and $C \geq 0$ is the commission size per order.

If during a strategy execution the suborders submitted by the strategy exhaust all limit orders offering to buy (or sell) at the spot price, the spot price will go down for *Sell* orders and up for *Buy* orders, and thus the *MSM* value will be positive. If the trader's suborders are executed with a price below the current mid-spread price for *Buy* and higher for *Sell* orders, then the *MSM* value will be negative and that strategy will be considered successful [26].

4.3   Naïve strategies

To estimate the performance of complex execution strategies, the main execution strategy types, such as those described in [13,25], can be used. Below we briefly summarize these strategies; for full details refer to [12,13,25,26].

- Strategies that submit only market orders:
  - Market Order Agent – submits 10 equal suborders at equal time intervals;
  - *VWAP* Agent – submits suborders at each tick, with $size = fullOrderSize/N$, where N is the number of tick updates received during the execution horizon H.
- Strategies that use limit orders:
  - Naïve Limit Order Agent – submits one limit order for the full quantity and amends the limit price to the current spot price at each tick;
  - Chunked Limit Order Agent – divides each order into 10 equal suborders, submits each chunk with equal intervals, and returns to chunks and amends them after a fixed time interval (less than the interval between the chunks' submissions).

4.4 Bayesian Adaptive Trading Agents

Theory. Bayesian adaptive trading with a daily cycle is a dynamic strategy for executing large orders [7]. This strategy is adjusted at each arrival of a price update (at each *price tick*). The derivation of the Bayesian-based strategy is based on the following assumptions [7]:

(i) The stock's price can be described as an arithmetic random walk: $S(t) = S_0 + \alpha t + \sigma B(t)$, where $S(t)$ is *the price of the stock at time t*, S_0 is *the price of the stock at the beginning of the day*, α is a *drift*, thought to be determined by the actions of the informed traders, σ is the market *volatility*, and $B(t)$ models *standard Brownian motion*.

Almgren *et al.* believe that large investors decide on their trading strategy in the morning and thus they determine the overall direction of the price movements [7]. This approach is in contrast with many execution strategies which treat price fluctuations caused by other traders as noise in the data [7]. The drift rate is assumed to be constant throughout the whole day, however, its value is unknown and has to be estimated via observations of the price movements. This assumption is supported by another assumption, that informed traders use strategies that have $VWAP$ as their benchmark. The volatility σ is determined from the actions of informed traders, as their trades are of a noise type and thus can be predicted on average [7]. Almgren *et al.* suggest to estimate σ from the short period of observations of price movements [7].

(ii) The true drift value is unknown, but it is assumed to be drawn from the Normal distribution.

$$\alpha \sim \mathcal{N}(\bar{\alpha}, \nu^2) \qquad (5)$$

This drift estimate will be updated as more stock price updates arrive during the day. According to Almgren *et al.* [7], the best estimate of α at time t is given by:

$$\alpha_\star(t) = \frac{\bar{\alpha}\sigma^2 + \nu^2(S(t) - S_0)}{\sigma^2 + \nu^2 t} \qquad (6)$$

(iii) It is assumed that the market impact of order execution can be described as a linear function of the execution speed.

The formulation of the problem is as follows: X represents the number of shares to be bought by the end of the trading horizon H, $x(t)$ represents the number of shares left to be bought at time t and is called *trading trajectory*. The trading rate (or speed of execution), $v(t)$, is defined as:

$$v(t) = -\frac{dx}{dt}; \quad \forall t \; v(t) \geq 0 \qquad (7)$$

The requirement of non-negativity means that the strategy should never sell to execute a *Buy* order.

Using assumption (iii), the price after the execution of a part of the order can be defined by:

$$\bar{S}(t) = S(t) + \eta v(t) \qquad (8)$$

where $\bar{S}(t)$ is the price of the stock after executing the strategy at moment t, $S(t)$ is the initial price of the stock at moment t, before any suborder was executed as a part of the strategy, and η is the market impact coefficient. For Buy orders $\eta \geq 0$, as every Buy order can only either drive the price up or leave it at the same level (according to the $Limit\ Order\ Book$ mechanism). Taking into account Eq. (7) and adding the requirement that the trading rate $v(t)$ should be continuous, Almgren $et\ al.$ [7] derived Eq. (9) for $x(\tau)$, where $t \leq \tau \leq H$:

$$x(\tau) = \begin{cases} \begin{cases} X & \tau < t^\star \\ \dfrac{H-\tau}{H-t}x(t) - \dfrac{\alpha_\star}{4\eta}(\tau - t)(H - \tau) & \tau \geq t^\star \end{cases} & \alpha_\star < -\alpha_c \\[3ex] \dfrac{H-\tau}{H-t}x(t) - \dfrac{\alpha_\star}{4\eta}(\tau - t)(H - \tau) & |\alpha_\star| \leq \alpha_c \quad (9) \\[3ex] \begin{cases} \dfrac{H^\star - \tau}{H^\star - t}x(t) - \dfrac{\alpha_\star}{4\eta}(\tau - t)(H^\star - \tau) & \tau \leq H^\star \\ 0 & \tau > H^\star \end{cases} & \alpha_\star > \alpha_c \end{cases}$$

$$t^\star = H - \sqrt{\frac{4\eta x(t)}{-\alpha^\star}}; \quad H^\star = t + \sqrt{\frac{4\eta x(t)}{\alpha^\star}} \quad (10)$$

α_c is the critical value of the drift, estimated using $\alpha_c = 4\eta x(t)/(H-t)^2$. Overall, the above equations mean that, if $\alpha_\star < -\alpha_c$, the trading does not start until the $t^\star$ period. If $\alpha_\star > \alpha_c$, then the trading finishes before the end period H – the last trade is expected to occur no later than at time $H^\star$. For exact details of the derivations refer to [7].

Extending the Standard Bayesian Agent Approach. In this section we propose three theoretical extensions to the standard Bayesian adaptive trading model presented in the previous section. We assume that trades will happen at discrete points of time $\{1, 2, \ldots, H\}$, time 0 is the beginning of the algorithm, and $x(0) = X$. The trade should be executed during one business day, so:

$$H = \frac{End\ of\ market\ day - beginning\ of\ market\ day}{tick\ step} - 1 \quad (11)$$

where $tick\ step$ indicates the time interval between two consecutive price updates ($e.g.$ every minute or 10 min).

(i) The maximum number of orders an agent can submit is H. Reference [25] mentions that the execution strategy may not consider the commissions, as execution strategies are usually employed by large institutional investors who receive considerably lower commission rates compared to the regular investors. However, it still seems appropriate to include the commissions into the calculation of the optimal strategy. If the algorithm trades too often, the commission incurred can be too high, as it is paid for every order executed. We can select H in such a way that the overall commission incurred for all orders will be no more than a certain percentage of the overall spot price of the order. We define the number of steps as:

$$step = \frac{H}{min(max(1, maxStepsNum), H)} \qquad (12)$$

where $maxStepsNum = S_0 X * \epsilon / ComSize$, ϵ is a commission threshold (e.g., 0.001) and $ComSize$ is the size of the commission per order. Then our strategy will act only when $t \bmod step \equiv 0$.

(ii) In [7] the schedule $x(\tau)$ (Eq. (9)) was derived for *Buy* orders. To adjust it to *Sell* orders, we analyse Eq. (8). Any *Buy* order can only worsen the spot price of a stock. For a *Sell* order, large orders will dive deeper into the *Limit Order Book*, and executing limit orders will gradually reduce prices the deeper the *Limit Order Book* is traversed, so the market impact equation becomes $\bar{S}(t) = S(t) - \eta v(t)$. Then the equation for the trading rate becomes:

$$x(\tau) = \frac{H - \tau}{H - t} x(t) + \frac{\alpha_\star}{4\eta}(\tau - t)(H - \tau) \qquad (13)$$

From here we can derive the *Sell* schedule $x(\tau)$:

$$x(\tau) = \begin{cases} \begin{cases} X & \tau < t^\star \\ \dfrac{H - \tau}{H - t} x(t) + \dfrac{\alpha_\star}{4\eta}(\tau - t)(H - \tau) & \tau \geq t^\star \end{cases} & \alpha_\star > \alpha_c \\[2em] \dfrac{H - \tau}{H - t} x(t) + \dfrac{\alpha_\star}{4\eta}(\tau - t)(H - \tau) & |\alpha_\star| \leq \alpha_c \\[2em] \begin{cases} \dfrac{H^\star + \tau}{H^\star - t} x(t) - \dfrac{\alpha_\star}{4\eta}(\tau - t)(H^\star - \tau) & \tau \leq H^\star \\ 0 & \tau > H^\star \end{cases} & \alpha_\star < -\alpha_c \end{cases} \qquad (14)$$

Further, by examining Eqs. (9) and (14) we can notice, that if in (14) we replace $\alpha^\star$ to $\alpha^{\star\prime} = -\alpha^\star$, then (14) becomes exactly (9). This means that the only difference in our algorithm for the *Sell* order is in the calculation of the drift estimate:

$$\alpha_\star(t) = \begin{cases} \dfrac{\bar{\alpha}\sigma^2 + \nu^2(S(t) - S_0)}{\sigma^2 + \nu^2 t} & \text{for a } \textit{Buy} \text{ order} \\[1.5em] -\dfrac{\bar{\alpha}\sigma^2 + \nu^2(S(t) - S_0)}{\sigma^2 + \nu^2 t} & \text{for a } \textit{Sell} \text{ order} \end{cases} \qquad (15)$$

(iii) How do we set $\bar{\alpha}$, ν^2 and σ for the current day in Eq. (6)? According to [7], if a trader believes that his prior belief $\bar{\alpha}$ is correct, then he needs to set the variance ν to 0; in this case price updates will not modify the value of α. If no reliable information about prior α is available, setting variance ν to ∞ will result in the estimate of the drift α to be formed only by price observations during the day. If we assume that the investors' overall drift for the previous day can be a good prior for the drift in the current day and that the absolute volatility σ is the same for both days, then the estimated value for the drift from the previous day can be evaluated using historical data for the previous day. Thus $S(t) = S_0 + \alpha t + \sigma B(t)$ and $S(t + 1) = S_0 + \alpha(t + 1) + \sigma B(t + 1)$. Subtracting one from another, we obtain:

$$S(t + 1) - S(t) = \alpha + \sigma(B(t + 1) - B(t)) \tag{16}$$

where S_0 is the stock price at the beginning of the trading period on the previous day, and $S(t)$ and $S(t + 1)$ are the stock prices at times t and $t + 1$, respectively, on the previous day. Since $B(t)$ models the Brownian motion, $B(t + 1) - B(t)$ is a random variable $\sim \mathcal{N}(0, t + 1 - t) = \mathcal{N}(0, 1)$ and $\sigma(B(t + 1) - B(t))$ can be considered a random variable drawn from $\sim \mathcal{N}(0, \sigma)$. Now if we use the MLE estimate [9]:

$$0 = \frac{1}{N - 1} \sum_{t=1}^{N-1} (S(t + 1) - S(t) - \alpha) \tag{17}$$

$$\sigma = \frac{1}{N - 1} \sum_{t=1}^{N-1} (S(t + 1) - S(t) - \alpha)^2 \tag{18}$$

where N is the number of price observations and $S(t)$ the stock price at time t on the previous day. From Eq. (17) we can derive:

$$\alpha = \frac{1}{N - 1} \sum_{t=1}^{N-1} (S(t + 1) - S(t)) \tag{19}$$

and then we can calculate σ using Eq. (18).

The assumption that the drift value from the previous day can be a good prior for the belief of the drift value of the current day means that *we believe that the intentions of the investors did not change much compared to the previous day*. The results of our experiments presented in Sect. 5 will show whether these assumptions lead to an improved performance of the trading algorithms, on backtesting and on market simulation.

5   Results

In this section we present the quantitative results and qualitative insights from the experiments we conducted with human-like and execution agent models implemented and assessed using the AlTraSimBa tool. Then we present the results of testing the Bayesian execution strategy and we compare it with naïve execution strategies. We also test whether the Bayesian agent assumptions we presented in Sect. 4.4 hold for backtesting and simulation.

5.1   Human-Like Agent Experiments

In this simulation model, human-like agents will be required to fill in a Limit Order book and simulate trade flows corresponding to real market trade flows. To assess whether the human-like agents generate a realistic flow of orders, from an economic perspective, we used two metrics frequently used in real markets – *the annualized daily returns volatility* and *bid-ask spread* [8,17,30].

Annualized daily returns volatility $(ADRV)$ is a common metric used to report the volatility of different stocks [16] and is calculated using the following equations:

$$Daily\ return = 100 * \frac{today\ closing\ price - yesterday\ closing\ price}{yesterday\ closing\ price} \quad (20)$$

$$ADRV = \sqrt{252} * standard\ deviation\ of\ daily\ returns\ for\ period \quad (21)$$

$$Spread = 100\frac{ask\ price - bid\ price}{\frac{ask\ price + bid\ price}{2}} \quad (22)$$

Reference [2] provides month-by-month statistics of annualized daily returns for 2893 stocks with different capitalizations. We downloaded these files and processed them in order to derive the "normal" levels of annualized volatility we would like our simulations to exhibit. The $ADRV$ mean and standard deviation results we obtained for the 2010 data are illustrated in Table 4. This analysis showed that 95 % of stocks had an annualized volatility between 1 and 80, and the interval of possible volatility values is [0.7250, 251.6504].

Table 4. Statistics of stocks ADRV for year 2010

Volatility	#	%	Example of stocks
0.7250–13.9316	2000	7.9261 %	[AON, CDNS, BCE]
13.9316–27.1382	7238	28.6847 %	[GXP, DBD, FLIR]
27.1382–40.3448	6953	27.5552 %	[BRF, IM, PPDI]
40.3448–53.5514	4472	17.7228 %	[WYNN, VNO, LEN]
53.5514–66.7580	2369	9.3885 %	[AEIS, EXK, HMIN]
66.7580–79.9646	1096	4.3435 %	[BGZ, GMO, MELA]
Annualized volatility mean: 37.4928			
Annualized volatility S.D: 22.2068			

Regarding the bid-ask spread, [22] suggests that for liquid stocks the average bid-ask spread is 2.0 %, and [11] reports average bid-ask spreads on NASDAQ and NYSE being 1.0133 % and 0.5773 % respectively, with 95 % of the stocks having spread of no more than 2.6486 % for NASDAQ and no more than 1.4442 % for NYSE. Thus, we concluded that, to validate our assumptions and modelling approach, our simulation experiments should produce price fluctuations with bid-ask spread no more than 3 % of the stock's mid-quote price.

In all our human-agents experiments we used a tick step of 1 min (every minute agents can submit new orders), a 6-hour trading day (from 13 to 19), a tick size of 1 cent; each market setting is run for 1 day to allow agents to fill the *Limit Order Book*. Simulations were run for a 10-day testing period, for which metrics are reported. We first calculated the average *Daily return* and then converted it to the annualized volatility using Eqs. (20) and (21). The *Spread* is calculated for all tick steps over the testing period using Eq. (22), and then the Annualized volatility mean and standard deviation are reported. The sizes of *Buy* and *Sell* queues are reported at the end of the testing period.

For each type of the market we ran several experiments with a different number of agents, to test whether a larger number of agents leads to more realistic results. For each market we reported *ADRV*, the bid-ask *Spread* mean and standard deviation, and the sizes of the *Buy* and *Sell* queues.

The conclusions of our analysis of the impact of different agent markets on the market validity are briefly presented next.

Markets with Random Trading Agents Only. In addition to the markets described in [21] (contrary, momentum and half-and-half) we decided to also investigate markets where all the agents are random trading agents (described in Sect. 2.2). This market type can be used if the no intelligence assumption is sufficient for the purpose of the research, and the experimenter believes that stock price fluctuations follow a random pattern.

The results of experiments for markets containing 5, 11, 22, 55 and 110 agents are presented in Table 5.

Table 5. Random agent market Annualized Volatility (*AV*) and bid-ask *Spread*

# Agents	AV	Spread mean	Spread S.D.	Buy queue	Sell queue
5	168.4439	3.8802	2.3468	1788	1584
11	73.0448	4.1120	2.5591	3801	4025
22	138.6540	4.1075	2.4528	7431	7761
55	109.9939	3.9487	2.3586	17905	17855
110	139.5403	3.7866	2.3630	33376	33430

From Table 5 we can notice that random trading agents fill the *Limit Order Book* quickly enough − after 1 day neither the *Buy* nor the *Sell* queue are ever empty (otherwise the *Spread* reported would have been *NaN*). Additionally, they fill limit books almost equally.

Daily returns for a different number of random agents are shown in Fig. 2. It appears that there is no pattern between the number of agents and daily returns fluctuations. That also explains why we cannot derive a pattern of *Annualized*

Volatility − *AV* seems to be independent of the number of agents. However, as reported in Table 4, *AV* is within (not very common) real-market values. Overall, markets generated by random trading agents can represent a market where a very volatile stock is traded.

By analysing the results, we saw that the *Spread* values for the 11-agent markets and for the 110-agent markets are the most volatile, however we see no reasonable explanation for this phenomenon. This behaviour is most likely caused by the underlying random nature of the agents used. The results in Table 5 show that the overall *Spread* mean and standard deviation are unchanged when increasing the number of agents. Spread values are slightly larger than expected for liquid stocks, because, as stated previously, for liquid stock we would expect spread values to be less than 3 %.

Overall, the market simulation with random agents seem to create an economically plausible situation, however slightly on the volatile side and more for illiquid stocks. The number of agents does not have a significant impact on the parameters of the market, and they should be chosen based only on volume (sizes of limit queues) a market should have.

Momentum Markets. Reference [21] in their experiments used a market with 50 momentum agents and 5 random agents. We additionally tested a *Llimit Order Book* filled by 10, 20 and 100 momentum agents, while maintaining the proportion of momentum and random agents. The summary of the experimental results is presented in Table 6.

Table 6. Momentum market Annualized Volatility (*AV*) and bid-ask *Spread* (The # Agents column contains the number of momentum : random agents.)

# Agents	AV	Spread mean	Spread S.D	Buy queue	Sell queue
10:1	118.1072	1.6752	1.2114	6512	7303
20:2	103.7858	1.4024	1.0369	14320	13483
50:5	149.2243	1.0546	0.7812	33613	34058
100:10	94.3928	0.8391	0.6262	59964	69918

For the momentum markets we observe again no connection between the Annualized Volatility and the number of the agents in the simulation model, however both *Spread* mean value and standard deviation decrease with increasing the overall number of agents. The *AV* values are plausible and comparable with the corresponding random agent market values. However the *Spread* values are within the target spread values of 3 %, and therefore suggest a more liquid stock. From Fig. 3 we can see that, indeed, for the 100:10 configuration market the *Spread* fluctuations are less volatile than for the markets with fewer agents. Both limit books in the momentum markets seem to be filled equally.

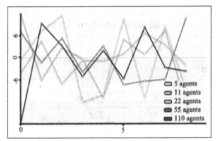

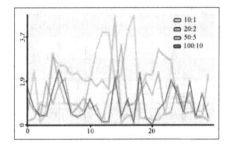

Fig. 2. Daily returns for random agents

Fig. 3. Bid-ask spread for the last 30 min of the momentum market

Table 7. Contrary market Annualized Volatility (*AV*) and bid-ask *Spread* (The # Agents column contains the number of contrary : random agents.)

# Agents	AV	Spread mean	Spread S.D	Buy queue	Sell queue
10:1	46017.1713	NaN	NaN	0	14749
20:2	478285.9925	NaN	NaN	16	22448
50:5	2063384.608	NaN	NaN	0	3666
100:1	1529582.742	NaN	NaN	1669	24768

Contrary Markets. Contrary markets according to [21] are the markets filled with contrary and random agents in proportion 10:1. The results we observed in experiments at the first sight seem perplexing (Table 7).

The values *NaN* in the columns for the bid-ask *Spread* standard deviation indicate that, during the testing period, one of the limit order books became empty. Indeed, we can see that the *Sell* queue is severely underfilled in comparison with the *Buy* queue. Additionally, the *AV* values indicate an extremely volatile stock, from an unrealistic (hopefully − unless another major crisis strikes) world. This is in accordance with [21], which, via experiments, showed that the contrary agents' actions quickly reverse any trend and simulate very volatile markets.

Even if such agents are allowed more time to fill the *Limit Order Book*, the antisymmetry feature of limit order books filling will result in one queue being periodically exhausted. Overall, it seems that contrary markets are not suitable for testing algorithmic trading strategies.

Half and Half Markets. Half and half markets are comprised of contrary, momentum and random agents in proportion 5:5:1, respectively [21] (Table 8).

The half-and-half market is more volatile than the momentum market, and the simulated stock is more illiquid (*Spread* of up to 7 %, with a high standard deviation). There is no observed connection between the number of agents and the *AV* and *Spread* values; both queues are filled approximately equally. This market type seems to be truly random, beating in randomness even the pure

Table 8. Half-and-half market Annualized Volatility (AV) and bid-ask *Spread* (The # Agents column contains the number of contrary : momentum : random agents)

# Agents	AV	Spread mean	Spread S.D	Buy queue	Sell queue
5:5:1	186.8334	4.3441	2.8278	3631	3847
10:10:2	260.533	7.1614	5.3655	7330	8087
25:25:5	236.0432	5.123	4.8293	16631	16636
50:50:10	187.67	3.5133	2.3815	32607	32855

random agent market. Indeed, according to the experiments carried out in [21], the half-and-half market was the least stable market, with larger price fluctuations.

Noise Markets. Reference [13] claims that, for noise agents with the parameters described in Table 1, the flow of orders generated is economically plausible and can be compared with the real market activity. Additionally, [18] has shown that zero intelligence agents with slightly different parameters could mimic real order flow in CDA simulations and are able to explain 96 % of the variance for bid-ask price fluctuations. Thus we would expect that zero intelligence agents can create a flow of orders which will result in realistic price and bid-spread fluctuations. However, it still remains a question of how many agents should we include into the market simulation and how many days should they first run before their overall activity simulates a realistic market setting.

Nevertheless, running the experiments with parameters from Table 1 we found out that the market thus generated is extremely volatile, and 1 full day is not enough to fill the *Limit Order Book* − over a 10-day testing period one of the limit books was exhausted each day, resulting in a *NaN* bid-ask *Spread*. We ran a sequence of experiments to see whether we can find better parameters for the agents. During testing it turned out that an extremely high probability of order cancellation (and only limit orders can be cancelled) in conjunction with a high market order probability resulted in queues being emptied very quickly by the arriving market orders. We ran a series of experiments with different probabilities for market, limit and order cancellation (where all probabilities ranging from 0.3 up to 0.9 were tried, with a 0.05 step) and the following different parameters were found to give more plausible results (Table 9) (different values were obtained only for the market order, limit order and order cancellation probabilities; the amended values are in red).

The results of the set of experiments with the parameters in Table 9 are summarized in Table 10. Interestingly, if we extend the period agents are given to fill the *Limit Order Book* before we test the market characteristics from 1 day to 5 days, the results obtained are economically more plausible, since annualized volatility is lower. Also, spread diminishes as more agents are used. This can be explained by more limit orders within the spread being entered, which diminishes the spread and increases the number of limit orders which offer to buy or sell at prices close to the spot prices, thus making incoming market orders execute

Table 9. The modified noise agent parameterization

Parameter	Value
Order cancellation probability	0.05
Market order probability	0.3
Market order size	$\sim powerlaw(2.7)$
Limit order probability	0.65
Limit order in spread probability	0.32
Limit order at spot price probability	0.33
Limit order outside the spread probability	0.35
Limit price inside the spread	$\sim uniform(bid, ask)$
Limit price outside the spread	$\sim powerlaw(2.5)$
Limit order size	$\sim powerlaw(2.1)$

Table 10. Noise markets Annualized Volatility (*AV*), bid-ask *Spread*, and *Buy* and *Sell* queues sizes. (Agents are given 1 and 5 days to fill the limit order book. The # Agents column contains the number of noise agents.)

# Agents	Annualized Volatility	Spread mean	Spread S.D	Buy	Sell
Agents are given **1** day before testing					
5	381.676	2.2934	6.5205	42423	408863
11	349.094	0.954	3.7674	65048	227519
22	492.48	0.1997	1.0485	223342	155818
55	27.8309	0.0339	0.3631	442579	851887
Agents are given **5** days before testing					
5	170.4793	0.6734	1.8316	55514	65256
11	161.0567	0.6092	1.7014	120106	109189
22	172.1393	0.4141	1.6055	211893	711861
55	64.3345	0.0836	0.5424	992644	922889

at closer prices. Overall, noise agents seem to fill queues unequally, but as time elapses (or more agents are used), the spread and volatility diminishes.

Human-Like Agents Markets Conclusion. Based on the results and analysis presented in Sect. 5.1, we concluded that the momentum market is the best model for testing algorithmic trading strategies, since it quickly fills the *Limit Order Book* and produces results comparable to those of a liquid stock. On the other hand, random markets can simulate a more volatile and less liquid stock, whereas markets containing noise agents should be used with more time (5 and more days) given to fill the *Limit Order Book*, and they can represent a very liquid and volatile stock.

5.2   Bayesian Adaptive Learners

Execution strategies aim at minimizing the market impact of the order to be executed, therefore they should be mainly tested on the simulation part of the software, since the market impact of each order could thus be measured. Backtesting assumes infinite liquidity and zero impact of any order submitted by an artificial agent, but it also considers all agents evenly, which cannot be achieved with the market simulation, due to the random nature of the simulation of price movements. So backtesting can still be used to check whether a strategy adjusts its aggressiveness of order submission as a consequence of price variations.

Experiment Setup. At first we filled the *Limit Order* books with the human-like agents of a specified market type, and saved the configuration to file. Then we ran each test on this initial configuration, to ensure the tests have the same initial conditions. We thus eliminated the situation where one of the strategies is launched on a more favourable market. Each agent was then given to execute a large order and statistics were calculated. In theory, the starting price for the security should be the same, since the market configuration is loaded from the configuration file for each test. However, due to the random order in which agents act, initial spot prices can differ slightly.

Simulations were run for five different configurations for two market types: momentum markets and random agent markets. For each agent, we reported the *MSM* metric with commissions considered (Eq. (4)), *VWAPR* (Eq. (3)), *the achieved price* of an overall execution and *the ideal price* of an order – the price of an order at its creation time.

For each saved initial market configuration, we estimated the daily trading volume. [13] claims that execution strategies are used to execute the orders of size 5–15 % of the Average Daily Volume (*ADV*). Orders less than 5 % are believed to have a small market impact and can be executed either as market orders or using simple executional strategies. Orders bigger than 15 % of *ADV* should be traded over several days.

For momentum markets tests, we used 50 momentum agents and 5 random agents, whereas for random markets we chose a configuration of 55 random agents. All markets were first run for 5 days before tests of different types of agents were launched.

Each Bayesian agent is initialised with two parameters: *Bayesian(consider-Commissions, learnPriors)*. The parameter *considerCommissions* indicates whether commissions should be considered when calculating the trading schedule. When it is set to *true*, the Bayesian agent might do less trading, in an attempt to minimize the commission costs; *learnPriors* indicates whether the agent should try to estimate σ and $\bar{\alpha}$ (Eq. (5) and Sect. 4.4) from previous days observations.

The simulation parameters we used are described in Table 11. We will consider the results for each market type separately and then will draw global conclusions.

Table 11. Simulation parameters for Bayesian agent experiments (Parameters marked with $*$ are derived from previous day observations if *learnPriors* is *true*. Parameters σ, $\bar{\alpha}$ and ν^2 are used to estimate the drift, using Eq. (15)).

Parameter	Value
Commission	8 units per order
Commission threshold	0.001 of initial order price
Tick step (updates every)	1 min
Market open time	13:00
Market close time	19:00
Default$*$ Market Volatility σ	1.5
Default$*$ prior estimate of drift $\bar{\alpha}$	0.7
Default prior STD of drift ν^2	1.0

Table 12. Bayesian versus naïve execution strategies for momentum markets. (The parameters correspond to *Bayesian(considerCommissions, learnPriors)*).

Order	Agent	Ideal	Achieved	MSM	MSM Rank	VWAPR
Buy	Bayesian (false,false)	564131.06	584106.45	354.0912	4	−9.1662
	Bayesian (true,false)	564139.06	594657.66	540.9765	7	96.6631
	Bayesian (false,true)	564131.06	581117.8	301.1133	3	449.5647
	Bayesian (true,true)	564139.06	589819.53	455.2152	6	321.1232
	VWAP	564131.06	589342.46	446.9067	5	39.5673
	NaïveLimitOrder	564131.06	575703.94	205.1452	2	−293.1889
	ChunkedLimitOrder	564131.06	560338.74	−67.224	1	−0.2298
Sell	Bayesian (false,false)	599083.8	559206.91	665.6312	7	173.0038
	Bayesian (true,false)	604936.52	568520.81	601.9757	5	304.0506
	Bayesian (false,true)	604928.52	565012.03	659.8546	6	332.0176
	Bayesian (true,true)	604936.52	565265.26	655.7921	4	−27.1817
	VWAP	608685.84	611498.17	−46.2033	1	−1.5198
	NaïveLimitOrder	608685.84	604873.3	62.6355	2	281.1482
	ChunkedLimitOrder	608685.84	588004.18	339.7756	3	104.0207

Momentum Market Results. Table 12 summarizes the results for the momentum markets. We remind the reader that for both *MSM* and *VWAPR*, the lower their values, the better is considered the strategy. We can see that estimating prior values for σ and $\bar{\alpha}$ for both *Buy* and *Sell* orders improves the *MSM* metric but decreases the *VWAPR* values. This indicates that estimating parameters from previous day observations provides a better starting point than using their default values. The improvement in *MSM* is more pronounced for the *Buy* order. The decrease in *VWAPR* is a side effect of the change in the trading speed, however the Bayesian strategy is not a VWAP-benchmark following strategy, since it bases its decisions only on sequences of price updates. Therefore poor *VWAPR* values in all types of Bayesian agents is an illustration that this strategy should not be used if a good *VWAPR* is needed. Not surprisingly, Bayesian agents receive the worst *VWAPR* in ranking when compared to naïve executional strategies (Table 12).

For *Sell* orders Bayesian agents (BA) that consider commissions and thus act less frequently than the standard BA improve the *MSM* only slightly, but deteriorate it for the *Buy* orders. This suggests that, even though such agents do less trading, they miss the important price changes updates that can result in an earlier adjustment of their execution strategies. This further suggests that perhaps setting a larger commission threshold may result in a bigger improvement of the BA performance, since they would receive more price updates and will have the opportunity to act more often. The trading trajectories (the sequence of the values of the number of shares an agent still has to buy or sell over time) for four Bayesian agents are presented in Fig. 4.

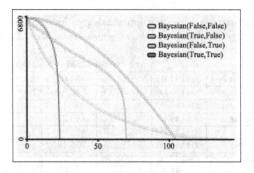

Fig. 4. The trading trajectories for Bayesian agents, in momentum markets, for *Buy* orders. The *x*-axis is the number of price updates, and the *y*-axis is the number of shares the agents still have to buy at each moment of time. The parameters correspond to *Bayesian(considerCommissions, learnPriors)*. (The agents that consider commissions are given fewer opportunities to act, and thus they are allowed to execute orders in a lower number of steps than the no-commission agents.)

We can see that the smoothest trajectory is of the agent that does not consider commission and uses default prior values (Bayesian(false,false)). Its trading speed increases slightly towards the end, when the agent trades less aggressively, thus minimizing the possibility of exhausting limit orders with prices close to the spot price and obtaining worse prices. Two agents which consider commissions have to buy a large number of shares at the end, since they did not acquire them during the trading, and this ought to drive the price in negative for an agent direction. This suggests that perhaps setting a larger commissions threshold may result in a bigger improvement of the Bayesian agents performance since they will receive more price updates and will thus have opportunity to act more often. The agent that estimates prior from previous day observations seems to have more even liquidation speed in comparison with the agent that uses default values and trades more slowly at the beginning.

Chunked limit order and naive limit order strategies outperform all Bayesian agents in terms of MSM, however this might be due to the fact that they engage in much less trading then even Bayesian agents that consider commissions. The trading trajectory for VWAP, naïve limit order and chunked limit order strategies are presented in Fig. 5. The VWAP agent steadily submits orders for equal quality at each price update it receives, whereas naïve limit order and chunked limit order agents execute orders much faster, thus they incur less commissions.

Random Market Results. Random market results and ranking of the agents according to the MSM metric are presented in Table 13.

Table 13. Bayesian versus naïve execution strategies and their MSM rank, for random markets. (The parameters correspond to $Bayesian(considerCommissions, learnPriors)$).

Order	Agent	Ideal	Achieved	MSM	Rank	VWAPR
Buy	Bayesian (false,false)	365764.64	380432.2	401.0109	6	346.1609
	Bayesian (true,false)	365772.64	369826.34	110.8256	2	660.0392
	Bayesian (false,true)	365764.64	385983.38	552.7801	7	522.92
	Bayesian (true,true)	365772.64	380246.36	395.7026	5	660.8258
	VWAP	365764.64	375378.25	262.8359	4	318.1245
	NaïveLimitOrder	365764.64	370027	116.5329	3	228.4315
	ChunkedLimitOrder	365764.64	361184.24	−125.228	1	−10.9637
Sell	Bayesian (false,false)	430498.36	417768.65	295.697	7	313.3508
	Bayesian (true,false)	417752.08	434243.34	−394.76188	4	730.0739
	Bayesian (false,true)	417744.08	411427.53	151.2062	6	437.2886
	Bayesian (true,true)	417752.08	411954.97	138.7691	5	292.4836
	VWAP	434805	482559.75	−1098.3026	1	278.2538
	NaïveLimitOrder	434805	466496.36	−728.8637	3	227.9997
	ChunkedLimitOrder	434805	475926.36	−945.7425	2	−131.7396

Estimating prior values in random markets considerably improves the MSM metric for the *Sell* orders, but deteriorates it slightly for the *Buy* orders. Since a random market is more volatile and less liquid than a momentum market, this suggests that estimates of priors based on previous days observations have less power in more volatile markets.

For both *Sell* and *Buy* orders the Bayesian agents that consider commissions considerably improve the MSM metric in comparison with the Bayesian agents that trade more frequently. However, it would be wrong to attribute such an improvement in MSM only to the influence of the commission parameter, as the market is volatile and less frequent trading can by chance lead to better results.

For *Sell* orders we can see that a simple strategy such as *VWAP* achieved an extremely good performance by trading very often. Naïve Limit Order and Chunked Limit Order strategies, just as in the Momentum market, outperformed all Bayesian agents, apart from the Bayesian agent with commissions, which

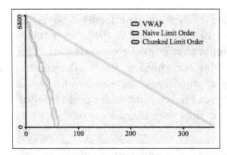
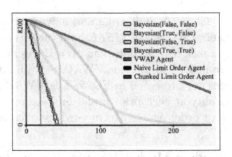

Fig. 5. Trading trajectory for simple executional strategies in momentum market for *Buy* orders. The *x*-axis is the number of price updates, and the *y*-axis is the number of shares agents still have to buy.

Fig. 6. Trading trajectory for executional strategies in random market for *Buy* orders. The *x*-axis is the number of price updates and actions of the agent, and the *y*-axis is the number of shares agents still have to buy.

was able to do slightly better than NaïveLimitOrder agents for a *Buy* order (Table 13).

The trading trajectories for *Buy* orders are presented in Fig. 6.

Noise Market Results. Noise market results are summarized in Tables 14 and 15.

This market is the most volatile and represents the least liquid stock. For this market considering commission estimates improved the performance only for the *Buy* orders, and significantly deteriorated performance for *Sell* orders. Prior parameters estimated did not bring any improved performance, which is explainable, since in such market trends do not last long enough and thus previous days values have no bearing on the current day price fluctuations. Chunked limit order and naïve limit order strategy again outperformed all the types of Bayesian agents.

BackTesting Results. We performed a backtest for a very liquid stock Amazon (AMZN) [5] for the period between 29 June−06 July 2010, with a one-minute tick step. We chose Amazon because it is actively traded and has large capitalization, thus it is possible that institutional investors will trade in this stock and that Almgren *et al.*'s assumption about the daily drift being formed by actions of the traders can hold for this data [7]. We can see in Fig. 7 that during this period there were two general trends - a downward trend from 25th June to 2nd July, followed by an upward trend until 9th July. Since backtesting does not allow to replicate a market impact of a large order, the size of the order does not matter, so we gave the agents the task to *Buy/Sell* 5000 shares. The results for 5 trading days for *Buy* and *Sell* orders are summarized in Table 16.

One of the problems in backtesting can be understood looking at the NaïveLimitOrder agent's performance indices. This agent submits a Limit Order for the full quantity at the current spot price. In real markets, if there is sufficient

Table 14. Bayesian agent results in comparison with naïve execution strategies for noise markets. (The parameters correspond to *Bayesian(considerCommissions, learn-Priors)*).

Order	Agent	Ideal	Achieved	MSM	VWAP Ratio
Buy	Bayesian(false,false)	790132.3	832538.73	536.7003	119.1093
	Bayesian(true,false)	790140.3	808998.53	238.6693	498.9715
	Bayesian(false,true)	790132.3	848776.9	742.2124	654.0119
	Bayesian(true,true)	790140.3	832450.65	535.4789	786.9869
	VWAP	790132.3	818957.2	364.8111	292.8517
	NaïveLimitOrder	790132.3	805074.84	189.1144	−148.7392
	ChunkedLimitOrder	790132.3	793131.89	37.9631	−118.1761
Sell	Bayesian(false,false)	509783.85	510380.53	−11.7046	470.7504
	Bayesian(true,false)	503590.1	497954.73	111.9039	426.3716
	Bayesian(false,true)	503582.1	501832.83	34.7365	435.5821
	Bayesian(true,true)	503590.1	506548.04	−58.737	165.9824
	VWAP	541536.81	591678.12	−925.9076	164.197
	NaïveLimitOrder	541536.81	576314.39	−642.2015	44.4981
	ChunkedLimitOrder	541536.81	576979.55	−654.4844	−29.5074

Table 15. Ranking of agents by the *MSM* metric for the noise market

Order	Rank	Agent	MSM
Buy (8264)	1	ChunkedLimitOrder	37.9631
	2	NaïveLimitOrder	189.1144
	3	Bayesian(true,false)	238.6693
	4	VWAP	364.8111
	5	Bayesian(true,true)	535.4789
	6	Bayesian(false,false)	536.7003
	7	Bayesian(false,true)	742.2124
Sell (8282)	1	VWAP	−925.9076
	2	ChunkedLimitOrder	−654.4844
	3	NaïveLimitOrder	−642.2015
	4	Bayesian(true,true)	−58.737
	5	Bayesian(false,false)	−11.7046
	6	Bayesian(false,true)	34.7365
	7	Bayesian(true,false)	111.9039

liquidity, an order is executed at the current spot price. However usually a Limit Order is only partially executed and the spot price goes down − for *Sell* orders, and up − for *Buy* orders. The remaining part of that Limit Order is amended subsequently to the spot price, at specified time intervals. Backtesting assumes unlimited liquidity, so the Limit Order is fully executed without modifying the price, as would happen in a real market. Actually, in backtesting, the Naïve execution strategy becomes the "Ideal (reference) strategy" that the *MSM* metric compares the agents' behaviour against.

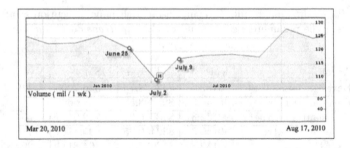

Fig. 7. Price fluctuations for AMZN stock (*Picture taken from Google Finance* [3])

Table 16. Backtesting: Bayesian agent versus naïve execution strategies for *Buy* (top) and *Sell* (bottom) orders. E_1 and SD_1 are the expectation and the standard deviation of MSM, respectively; E_2 and SD_2 are the expectation and the standard deviation of $VWAPR$, respectively.

Agent	29/6	30/6	1/7	2/7	5/7	E_1	SD_1	E_2	SD_2
Bayesian(false,false)	−9.75	202.88	−1.27	−115.97	−0.04	15.16	103.49	15.28	86.04
Bayesian(true,false)	−8.45	258.94	86.02	−150.45	12.06	39.62	133.75	29.53	69.36
Bayesian(false,true)	−9.75	279.89	100.22	−167.26	−0.04	40.61	147.11	40.03	66.52
Bayesian(true,true)	−8.45	257.72	118.67	−169.77	12.06	42.04	141.91	33.30	68.92
VWAP	−9.84	273.91	32.15	−144.35	0.02	30.37	135.93	30.02	74.75
NaïveLimitOrder	0	0	0	0	0	0	0	1.43	151.87
ChunkedLimitOrder	−9.84	258.45	25.49	−134.51	109	49.71	130.50	27.67	75.79
Bayesian(false,false)	9.93	−264.39	−47.80	155.82	−0.03	−29.29	135.85	−28.91	72.32
Bayesian(true,false)	8.63	−227.31	−105.28	162.86	−9.62	−34.14	129.32	−26.12	71.21
Bayesian(false,true)	9.93	−223.08	52.56	93.77	−0.03	−13.36	110.02	−13.54	100.02
Bayesian(true,true)	8.63	−230.64	−63.87	138.38	−9.62	−31.42	119.71	−22.55	73.36
VWAP	9.84	−273.91	−32.15	144.35	−0.02	−30.38	135.93	30.02	74.75
NaïveLimitOrder	0	0	0	0	0	0	0	1.435	151.87
ChunkedLimitOrder	9.84	−258.45	−25.49	134.51	0	−27.91	127.86	27.671	75.79

Another problem is revealed when comparing metrics for *Buy* and *Sell* orders (Table 16). Generally speaking, if the price of a security goes down, the best *Buy* strategy is the one that buys closer to the end of the period. If the price goes up, then any *Sell* order strategy that breaks an order into chunks will benefit. This is why we can observe in backtesting that, if an agent has a negative *MSM* value for a *Buy* order, it will have a positive *MSM* value for a *Sell* order on this day. Further, simpler strategies like Naïve Limit order, VWAP agent and Chunked Limit Order have exactly opposite in sign values for *Buy* and *Sell* orders, since they do not adjust their behaviour based on the order's direction.

Considering commissions for Bayesian agents deteriorates the *MSM* metric for both *Buy* and *Sell* orders on all days apart from 5th July for the *Sell* order and 2nd July for the *Buy* order (Table 16). Actually, on the trading days before 5th July the closing price was less than the closing price of the previous day, the trend was downwards, as we can see in Fig. 7, and on 5th July the trend was already upwards. So it seems that when the trend is upwards, the option to consider commissions results in an improvement of the *MSM* metric. Overall, less frequent trading (the one we obtain when considering commissions) should be favourable for *Buy* orders – when the price trend is downwards – and for *Sell* orders – when the price trend is upwards. However, daily fluctuations of prices can result in this rule not working, as we can see for other days for *Buy* and *Sell* orders (Table 16).

Estimating priors increases the performance of an agent only on 2nd July, for both *Buy* and *Sell* orders. This is probably due to the fact that the price fluctuations of the 1st and 2nd of July are indeed correlated. Interestingly, on the 1st July the news "Amazon bought wacky retailer" were announced [1], which probably resulted in investors selling stocks over 1st and 2nd July, which then created correlated overall drifts in price.

Finally, if we do not consider Naïve Limit order strategy which leads to a 0 *MSM* metric value due to backtest infinite liquidity, we can notice that the Bayesian agents obtain the best *MSM* mean value for the period considered.

Bayesian Agent Conclusions. Bayesian agents assume that market price fluctuations are defined by the bulk number of intelligent institutional investors which determine the overall direction of price movements, thus the market created by such investors is not supposed to be very volatile. Estimating priors ($\overline{\alpha}$ and σ) of Bayesian agents based on previous day observations is advantageous only in stable markets or if there is a strong correlation between price fluctuations for two days, for example when these days are close to the day a financial report was released or stock news were announced.

Considering commissions may result in large orders being executed at the end of the period and thus in a deterioration in the *MSM* metric values. This could perhaps be mitigated by increasing the commission threshold, so the agents trade more frequently. The more frequent trading seems to improve the overall performance since, even though the agents encounter higher commissions, they are given more frequently the opportunity to react to price movements and can therefore adjust their aggressiveness at the earliest possible moment. Finally, as Bayesian agents are not focused on achieving good VWAP Ratio values, they should not be used to model a VWAP-achieving strategy market.

We would like to conclude by noting that the core assumption of the Bayesian agents is that there is a daily drift in prices caused by the investors' intelligent actions. Almgren *et al.* in [7] mention that this assumption cannot be proven by empirical studies, but it probably does take place, since some real traders express interest in this strategy. From our observations, it seems that, for real data, this assumption does not necessarily hold each day, otherwise Bayesian agents would have consistently outperformed other strategies in the backtesting configuration.

6 Conclusions and Future Work

On the basis of the results presented in this paper, we propose that *momentum market* is the most suitable model for testing algorithmic trading strategies, since it quickly fills the Limit Order book and produces results comparable to those of a liquid stock (Sect. 5.1). The standard Bayesian execution agent model

(Sect. 4.4) was implemented and its performance assessed. Additionally, we proposed an extended Bayesian model that models commissions when generating the schedule of order executions, and also models the ability to estimate prior values for the daily price drift and market volatility, based on previous day observations. The schedule for *Sell* orders was analytically derived using a methodology similar with that employed in [7] for *Buy* orders. Our experiments revealed that the method of estimating priors proposed in this paper can be advantageous in relatively stable markets, when trading patterns in consecutive days are strongly correlated. The results also showed that adjusting the execution strategy according to a reducing commission criterion can result in a worse performance, since agents trading less frequently seem to miss important price trend changes and act with a delay. Additional experiment results and insight analysis, which could not be included in this paper due to space constraints, are included in [27].

In terms of future work, in the Bayesian agent analysis it would be worth experimenting with different commission thresholds, to find the best proportion of price updates frequencies and trading.

Acknowledgements. Natalia Ponomareva would like to gratefully acknowledge the Hill Foundation for supporting her study for an MSc degree at the Department of Computer Science of the University of Oxford.

References

1. Amazon buys wacky retailer. http://www.theregister.co.uk/2010/07/01/amazon_buys_woot/
2. Cboe historical stock volatilities. http://www.cboe.com/data/historicalvolatility.aspx
3. Google finance. http://www.google.com/finance
4. JASA - Java Auction Simulator API. http://www.essex.ac.uk/ccfea/research/software/jasa/
5. Online tick-level dataset Dukascopy. http://freeserv.dukascopy.com/exp/
6. A resource for agent- and individual-based modellers, and the home page of Swarm. http://www.swarm.org/index.php/Main_Page
7. Almgren, R., Lorenz, J.: Bayesian adaptive trading with a daily cycle. J. Trading **1**(4), 38–46 (2006)
8. Andersen, T.G., Bollerslev, T., Diebold, F.X., Ebens, H.: The distribution of realized stock return volatility. J. Financ. Econ. **61**, 43–76 (2001)
9. Bishop, C.M.: Pattern Recognition and Machine Learning. Springer, Heidelberg (2006)
10. Bonabeau, E.: Agent-based modeling: methods and techniques for simulating human systems. PNAS **99**(3), 7280–7287 (2002)
11. Chung, K.H., Kim, Y.: Volatility, market structure, and the bid-ask spread. Asia-Pac. J. Financ. Stud. **38**(1), 67–107 (2009)
12. Cui, W., Brabazon, A., O'Neill, M.: Efficient trade execution using a genetic algorithm in an order book based artificial stock market. In: Conference Companion on Genetic and Evolutionary Computation, pp. 2023–2028 (2009)
13. Cui, W., Brabazon, A., O'Neill, M.: Evolving dynamic trade execution strategies using grammatical evolution. In: Di Chio, C., et al. (eds.) EvoApplications 2010, Part II. LNCS, vol. 6025, pp. 192–201. Springer, Heidelberg (2010)

14. Daniel, G.: Asynchronous simulations of a limit order book. Ph.D. thesis, University of Manchester (2007)
15. Domowitz, I., Yegerman, H.: The cost of algorithmic trading: a first look at comparative performance. J. Trading **1**, 33–42 (2006)
16. Ederington, L.H., Guan, W.: Measuring historical volatility. J. Appl. Financ. **16**, 5–14 (2006)
17. Engle, R., Patton, A.: What good is a volatility model? Quant. Financ. **1**, 237–245 (2001)
18. Farmer, J.D., Patelli, P., Zovko, I.I.: The predictive power of zero intelligence in financial markets. SSRN eLibrary (2004)
19. Gsell, M.: Assessing the impact of algorithmic trading on markets: a simulation approach. In: 16th European Conference on Information Systems, pp. 587–598 (2008)
20. Hendershott, T., Jones, C., Menkveld, A.: Does algorithmic trading improve liquidity? J. Financ. **66**, 1–33 (2011)
21. Izumi, K., Toriumi, F., Matsui, H.: Evaluation of automated-trading strategies using an artificial market. Neurocomputing **72**, 3469–3476 (2009)
22. Jiang, C.X., Kim, J.-C., Wood, R.A.: A comparison of volatility and bidask spread for NASDAQ and NYSE after decimalization. Appl. Econ. **43**(10), 1227–1239 (2011)
23. Kakade, S.M., Kearns, M., Mansour, Y., Ortiz, L.E.: Competitive algorithms for VWAP and limit order trading. In: Proceedings of the 5th ACM Conference on Electronic Commerce, pp. 189–198 (2004)
24. Lebaron, B.: Agent-based computational finance. In: Handbook of Computational Economics, Agent-based Computational Economics, pp. 166–209 (2006)
25. Nevmyvaka, Y., Feng, Y., Kearns, M.: Reinforcement learning for optimized trade execution. In: Proceedings of the 23rd International Conference on Machine Learning, pp. 673–680 (2006)
26. Nevmyvaka, Y., Kearns, M.S., Papandreou, A., Sycara, K.P.: Electronic trading in order-driven markets: efficient execution. In: CEC'05, pp. 190–197 (2005)
27. Ponomareva, N.: Using agent-based modelling and backtest to evaluate algorithmic trading strategies. Master's thesis, Department of Computer Science, University of Oxford (2011)
28. Raghavendra, S., Paraschiv, D., Vasiliu, L.: A framework for testing algorithmic trading strategies. Working Paper No. 0139 (2008). http://hdl.handle.net/10379/325
29. Rashid, A.: Using a service oriented architecture for simulating algorithmic trading strategies. In: Proceedings of the 12th International Conference on Information Integration and Web-based Applications & #38; Services, iiWAS '10, France, Paris, pp. 925–929 (2010)
30. Roll, R.: A simple implicit measure of the effective bid-ask spread in an efficient market. J. Financ. **39**, 1127–1139 (1984)
31. Sharpe, W., Alexander, G.J., Bailey, J.W.: Investments, 6th edn. Prentice Hall, New York (1998)
32. Wang, F., Dong, K., Deng, X.: Algorithmic trading system: design and applications. Front. Comput. Sci. China **3**, 235–246 (2009)
33. Wurman, P.R., Walsh, W.E., Wellman, M.P.: Flexible double auctions for electronic commerce: theory and implementation. Decis. Support Syst. **24**(1), 17–27 (1998)

Self-Explanation in Adaptive Systems
Based on Runtime Goal-Based Models

Kris Welsh[1](✉), Nelly Bencomo[2], Pete Sawyer[3], and Jon Whittle[3]

[1] School of Computing, University of Kent, Canterbury, UK
k.welsh@kent.ac.uk
[2] School of Engineering and Applied Science, Aston University, Birmingham, UK
nelly@acm.org
[3] School of Computing and Communications, Lancaster University, Lancaster, UK
{sawyer,whittle}@comp.lancs.ac.uk

Abstract. The behaviour of self adaptive systems can be emergent, which means that the system's behaviour may be seen as unexpected by its customers and its developers. Therefore, a self-adaptive system needs to garner confidence in its customers and it also needs to resolve any surprise on the part of the developer during testing and maintenance. We believe that these two functions can only be achieved if a self-adaptive system is also capable of self-explanation. We argue a self-adaptive system's behaviour needs to be explained in terms of satisfaction of its requirements. Since self-adaptive system requirements may themselves be emergent, we propose the use of goal-based requirements models at runtime to offer self-explanation of how a system is meeting its requirements. We demonstrate the analysis of run-time requirements models to yield a self-explanation codified in a domain specific language, and discuss possible future work.

Keywords: Self-explanation · Self-adaptive · Goals · Claims

1 Introduction

Self-adaptive systems are able to adjust their behaviour according to changes in their operating environment. Uncertainty in the operating environment may cause the behaviour of self-adaptive systems to be emergent. A system whose behaviour cannot be accurately predicted poses serious problems in terms of assurance and acceptance. A lack of intelligibility may cause users to stop using a self-adaptive system [1–3]. Because its behaviour is emergent, a self-adaptive system needs to garner confidence in its stakeholders, and allow developers to

---

This paper is an extended version of the paper "Towards Requirements Aware Systems: Run-time Resolution of Design-time Assumptions" by Bencomo, Welsh, Sawyer and Whittle, 17th IEEE International Conference International Conference on Engineering of Complex Computer Systems ICECCS 2012, Paris, France, July 2012.

R. Kowalczyk and N.T. Nguyen (Eds.): TCCI XVI, LNCS 8780, pp. 122–145, 2014.
DOI: 10.1007/978-3-662-44871-7_5

understand observed behaviour [3]. We believe that these two functions can only be achieved if a self-adaptive system is also capable of self-explanation.

We argue that a self-adaptive system's behaviour is best explained in terms of the satisfaction of its requirements. Observing the degree to which a system satisfies its requirements is well-discussed in requirements monitoring literature [4], and addresses questions of *what* the system is doing. The ability of a self-adaptive system to select alternative configurations based on environmental triggers raises questions on *how* the system is doing it, with more useful explanations offering clues to *why* the system is behaving as observed. Readily-understandable explanations are challenging to produce, with several key challenges preventing developers from creating such functionality. These challenges are discussed in the following paragraphs.

Firstly, an ability to explain behaviour relies upon an ability to monitor, introspect and reason about the system's current and past behaviour. There has been significant research interest in providing support for requirements monitoring [4,5], and in the specific area of self-adaptive systems, advances have also been made towards better support for introspection by adaptive middleware [6,7] and other frameworks [8,9]. However, work seeking to combine these two capabilities with reasoning still needs more research effort. The new and broader research area of requirements-aware systems covers similar interests [3,10,11].

Secondly, explanations need to be created at a sufficiently high level as to be understandable by a variety of interested stakeholders (e.g. end-users, but also by maintainers and support personnel). Ideally, users should interact with the system at a level of abstraction that is meaningful to them. This requires that the system is able to trace backwards and forwards between abstractions at the user's level and abstractions used by the systems at lower levels (e.g. components, component configurations, etc.). Furthermore, a trace of relevant events in the history of the adaptations the system has gone through should be kept by the system.

Thirdly, for self-explanations to be trustable, a self-adaptive system should be able to trace down from goals towards code to keep a synchronized link between requirements and architecture during execution. This trace needs to consider the dynamic changes that will affect requirements and the architecture of the system at runtime and keep a causal connection between the two.

Finally, a self-adaptive system should be able to reproduce a trace history of the adaptations it has performed in a way that is meaningful to support self-explanation.

In [12], we described our view of requirements-aware systems. In our work, representations of assumptions are made explicit using the concept of claims in goal models at design time. Using what we call claim refinement models (CRMs), we have defined the semantics for claims in terms of their impact on alternative strategies that can be used to pursue the goals of the system. The impact is calculated in terms of satisfaction and trade-off of the system's non-functional requirements (modeled as softgoals). Crucially, at runtime, when the executing system monitors that a given claim does not hold anymore, the system may adapt to an alternative goal realization strategy that may be more suitable for the new contextual conditions. Importantly, our approach tackles uncertainty, i.e. the new goal realization strategy may imply a new configuration of components that was not necessarily foreseen at design time. With the potential for unforeseen behavior, self-explanation capabilities are crucial. In this paper we build on the approach described in [12] to address the challenges posed by self-explanation described above.

The rest of the paper is organized as follows: In Sect. 2 we present the motivation of the paper using a simple but yet useful discussion. In Sect. 3, we discuss our initial progress towards a mechanism by which self-explanation can be achieved. In Sect. 4, we apply this means of providing self-explanation to a short case study. In Sect. 5, we propose a simple domain-specific language in which to convey self-explanations generated using our technique. Section 6 describes relevant related work. Section 7 concludes the paper and discusses future work.

2   Motivating Example

Consider the example of a robotic vacuum cleaner for domestic apartments, which uses self-adaptation to balance two conflicting non-functional requirements: to avoid causing a danger to people within the apartment (*avoid tripping hazard*) and to be economical to run (*minimise energy costs*). The cleaner supports two modes of operation: clean at night and clean when empty. Cleaning at night will likely yield lower energy costs, but could cause the occupants to trip should they awake and move about the apartment. Cleaning when the apartment is empty eliminates this hazard, but if the apartment is only empty during daytime this will come at a cost of increased energy costs. A standard goal model, showing the different ways in which the robot can clean the apartment, and each method's impact on the two competing NFRs (which can be modelled as softgoals) would be deadlocked, with no clear favourable goal operationalisation strategy. We have previously discussed [13] the use of claims, which were first proposed in the Non-functional Requirements (NFR) Framework [14], to model an assumption made to break the deadlock in a goal model. In this case, we can make an assumption that the tripping hazard is unlikely to cause an accident. We illustrate this using an i* [15] Strategic Rationale (SR) model, which models how an agent achieves its goals, and allows alternative goal satisfaction strategies to be compared in terms of their impact on softgoals. The model in Fig. 1 shows a claim "No Tripping Hazard" breaking the deadlock that would otherwise occur.

In Fig. 1, the vacuum cleaner's "Clean Apartment" goal may be satisfied either by the "Clean at night" task, or the "Clean when empty" task. Cleaning at night *helps* satisfy the "Minimise energy costs" softgoal, but *hurts* the "Avoid tripping hazard" softgoal, as represented by the *contribution links* attached to the task. The "Clean when empty" task makes the inverse contributions to each of the softgoals.

The "No tripping hazard" claim *breaks* the negative contribution made to the "Avoid tripping hazard" softgoal by the "Clean at night" task, which means

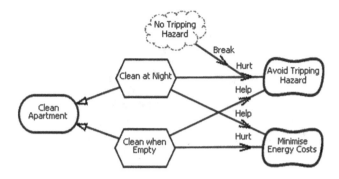

Fig. 1. Goal model of a robot vacuum cleaner from [16]

that this contribution should be lent less credence, or disregarded completely when deciding between the competing goal operationalisation strategies. With this assumption made, the decision to clean at night follows naturally.

Although assuming that the tripping hazard doesn't pose any real risk makes for a convenient way to break the deadlock in the goal model, the assumption is mere conjecture and would prove difficult to verify at design time. Thus, the robot vacuum cleaner is provided with a means of verifying the assumption at runtime, using monitoring. The broad nature of the "No tripping hazard" claim makes it more difficult to identify a suitable monitoring mechanism, so we use a *claim refinement model* (CRM) to decompose the claim hierarchically into its underlying assumptions, until some more precise, and crucially monitorable, assumptions are identified. We consider a claim refinement model to be sufficiently complete when all leaf claims are either: monitorable, axiomatic or considered an unmitigatable risk. In the latter case, the claim marks the edge of the contextual envelope in which the system is capable of tailoring itself to suit.

In this example, our "No tripping hazard" claim can be decomposed into the CRM shown in Fig. 2. There are four sub-claims organized in two ANDed branches (claims may also be OR-ed). Together, the branches illustrate the rationale for why the root claim should hold. In this case, "No tripping hazard" holds because there is no-one in the room in which the vacuum cleaner is working AND no external impact has been detected by the vacuum cleaner. The leaf claims of the CRM, "Light level [remains] constant" and "No shock detected" are directly monitorable via events or statistical data collected by the system. We refer to claims that are possible to directly monitor and verify at runtime as *monitorables*. If a monitorable turns out to be false, for example, if the vacuum's inertial sensor detects an external shock, then claim falsification propagates upwards towards the root. Thus, in this case, the impact event would falsify the "No tripping hazard" claim by propagation. Similarly, a sudden increase in the light level would indicate that a light has been switched on by a woken occupant, and the "No Tripping Hazard" claim would again be falsified by propagation.

With a means of run time verification for the deadlock-breaking "No tripping hazard" assumption having being found, the robot vacuum cleaner can be

Fig. 2. Claim refinement model for robot vacuum cleaner

specified as using a clean at night strategy unless a shock is detected or a light is switched on, in which case the robot should self-adapt to use the "Clean when empty" strategy.

However, after it has been in operation for some time, the owners of the robot vacuum cleaner find that it is costing more to run than expected. A self-explanation capability would mean that the vacuum cleaner could explain that it is required to avoid causing a tripping hazard, and that it has been unable to clean at night because the occupants frequently wake up and turn the lights on. In this scenario, the explanation would help the users to understand the system's behaviour, and help to pinpoint the reason the system is not behaving in the manner they would have imagined. The customer understands the reasoning, but is still dissatisfied because the operating costs are unacceptable. They submit a change request to the developer for the vacuum cleaner to be modified so that it only adopts the clean when empty strategy if two consecutive nights' cleaning have been interrupted.

In isolation, this change request may seem unimportant to the developers, especially if the change request is scant on background information justifying it. To contextualize the request, they interrogate the vacuum cleaner to determine its history of operation, with special attention to its history of self-adaptation and the events sensed in its environment that triggered adaptations. They discover the light detection event is being triggered more frequently than expected, and understand by consultation of the requirements model that this is interpreted as invalidation of the assumption that underpins prioritization of energy cost minimization.

The developers realize that running costs are high but note also that the customer does move around the apartment at night. They modify the vacuum cleaner's software to adopt a new strategy; they relax [17] the clean apartment goal by accepting that the clean apartment goal may be satisfied at a later time. The user change request is accepted; when interrupted, the robot tries to clean the following night before resorting to the clean when empty strategy.

In this simple example, the information contained within the explanation offered by the system could be obtained by analysis of standard debugging output or logs, and by deduction. However, these sources of information are low-level artefacts of particular code execution paths, and such analysis is performed by the system's developers, who will need time to perform the analysis. The potential for a self-adaptive system to adopt an unexpected configuration, or adopt an expected configuration in unexpected circumstances, means that there is a need for users to be able to understand what the system is doing, and why.

Our interest lies in reconciling a higher-level trace of the system's behaviour with its requirements, to establish whether the system's behaviour is appropriate, or better optimal, and whether the requirements themselves are correct. Although an explanation in terms of requirements may still prove too complex for some users to be able to understand a system's operation in some circumstances, the higher-level explanation may allow non-developer support personel to resolve queries without requiring developer input.

3 Self-Explanation Through Run-Time Requirements Models

Andersson *et al.* propose a means of characterising the change a self-adaptive system is designed to tolerate. Changes can be *foreseen, foreseeable* or *unforeseen*, as explained in [18]. We ignore here systems dealing with unforeseen change, which are more properly a topic for artificial intelligence research and pose a different order of challenge both for self-adaptation and self-explanation.

Much of our previous work has concerned requirements modeling for systems dealing with *foreseen* change [13,16,19]. Where change is foreseen, the set of contexts that the system may encounter are known at design time. Here, a self-adaptive system can be defined as a set of pre-determined system configurations that define the system's behaviour in response to changes of environmental context. Thus, there is little or no uncertainty about the nature of the system's environment and, if it is developed to high quality standards, satisfaction of the systems requirements should be deterministic.

More recently [12], we have started to address systems dealing with change that is, in [18]'s terms, merely *foreseeable*. Here, the key challenge is uncertainty, where at design time some features of the problem domain are unknown, perhaps even unknow*able*. Crucially, and in contrast to unforeseeable change, the fact of this uncertainty can be recognized, offering the possibility of mitigating it by resolving the uncertainty at runtime. The uncertainty associated with foreseeable change typically forces the developers to make assumptions in order to define the means to achieve the system's requirements. Thus, for example, a particular environmental context may be assumed to have particular characteristics and the system's behaviour defined accordingly. If the context turns out to have different characteristics, the system may behave in a way that is inappropriate. This has led us to exploit the concept of markers of uncertainty. Markers of uncertainty serve as an explicit marker of an unknown that forces the developer to make an assumption. We implement markers of uncertainty using claims as described in the previous section. A benefit of using claims to represent design-time assumptions is that the uncertainty is bounded and thus the risk of the system behaving in an inappropriate may be mitigated by monitoring, claim and goal evaluation, and adaptation.

Our solution uses i* goal and claim refinement models, as depicted in Figs. 1 and 2. As described in the previous section, claim monitoring may permit assumptions to be verified during operation. Where a claim turns out to be false, the corresponding portion of the goal model can be re-evaluated at run-time with the claim removed, or its effect on the model weakened. If, as a consequence of this, the original goal operationalisation strategy no longer evaluates as the optimal solution, an alternative goal operationalisation strategy can be substituted dynamically, using the system's adaptation mechanism. We have applied our work to the domain of wireless sensor networks where our run-time models are supported by advanced adaptive middleware and domain-specific component models [6]. The overview of the approach is shown in Fig. 3. The overview is explained in terms of the development process (the box @design time) and the run-time components (the box @runtime). The module *Self-Explanation* is part of the module *Runtime Reasoner*, which is responsible for the transformation of the run-time goal models, as will be explained further in the next section.

In the context this paper, the key feature of foreseeable change is that it may result in behaviour that is emergent. Emergent behaviour may surprise stakeholders who may require the behaviour to be explained in order to build and maintain their confidence in the system. Our thesis is that the same run-time requirements models that we employ to handle unforeseen change can also be employed as the basis of a self-explanation capability. Partially based on [20], a useful self-explanation of an adaptation needs to include:

1. Details of any change in priority, or the proposed degree of satisfaction of a system (soft)goal.
2. Details of the adaptation performed by the system.
3. The history of the adaptation, and the related events that triggered it.

In the next section, we illustrate how a self-explanation of a system's behaviour that contains this information may be provided using our run-time requirements models solution for the GridStix wireless sensor network.

4   Case Study

To demonstrate self-explanation in the context of a system which adapts to contexts not fully foreseen, we present the GridStix flood prediction system [21]. We have previously discussed this system in the context of requirements modelling [13], and have recently been exploring run-time uses of these requirements models. In [12], we discuss systems using run-time goal-based models to guide adaptation to circumstances where assumptions on which the originally prescribed configuration(s) rely no longer hold. In this paper, we show how claims and run-time requirements models that have been implemented for GridStix support self-explanation.

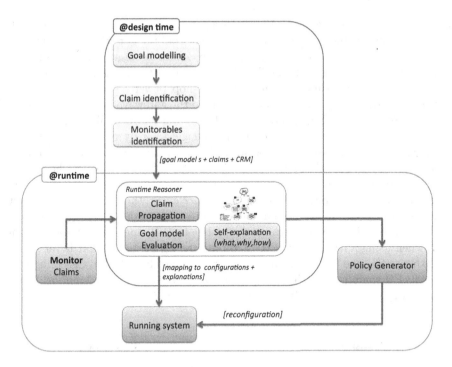

Fig. 3. Overview of the approach

The GridStix system is a wireless sensor network (WSN) for detecting and predicting flooding, versions of which were deployed on the river Ribble in North-West England and on the River Dee in North Wales. GridStix comprises a number of nodes (14 on the Ribble installation), each of which are equipped with sensors for detecting water depth and flow rate. The captured sensor data is processed by a stochastic model of the river to predict future river state. A feature of this algorithm is that it is distributed and lightweight enough to be executable by the GridStix nodes. Incremental results are cascaded from the most up-stream node down to the gateway node and from there via a GSM link to Lancaster University. Its accuracy is a function of the number of nodes contributing data.

GridStix is deployed in relatively remote, inaccessible locations with no mains power available, requiring that GridStix nodes rely on batteries and solar panels for power. As a result, energy conservation is a key non-functional requirement. GridStix uses an ad-hoc overlay network in which nodes can communicate using Bluetooth or WiFi, configured as either a shortest-path or fewest-hop spanning tree.

To help test feasibility and derive requirements for GridStix, empirical data was collected from experiments with a laboratory-based prototype. Data was collected to measure (among other metrics) resilience and power consumption [6], as illustrated by the graphs in Fig. 4. Here, resilience is a measure of network fragmentation; the more nodes become isolated from the gateway (uplink) node, the less resilient is the network. If too many nodes become isolated from the gateway node, it becomes impossible for the system to offer an accurate flood prediction. Power consumption measures per-hop power consumed during the transmission of 1 KB of data from each node to the gateway. The graph *Physical Network Resilience* in Fig. 4 shows that the greater range of WiFi meant that data from each node could be routed to the gateway by a larger number of paths with WiFi than using Bluetooth, while the graph *Physical Network Power Consumption* in Fig. 4 shows that the additional resilience comes at the cost of higher power consumption.

Similarly, the graph *Spanning Tree Resilience* shows that, for a small number of nodes (nodes B, H and I), the number of routes to the gateway affected by node failure is much higher when using a shortest-path (SP) spanning tree algorithm than when using a fewest-hop (FH) spanning tree. This means that fewer nodes are likely to become isolated from the gateway node when GridStix is configured to use its FH spanning tree. The graph *Spanning Tree Power Consumption* shows that for the nodes furthest from the gateway node (nodes L, M, N and O) the

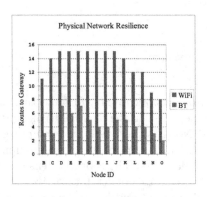
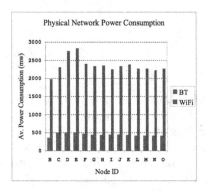
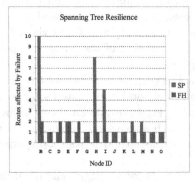
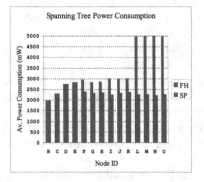

Fig. 4. Laboratory performance data (reproduced from [6])

power consumed in transmitting the data is significantly higher for a FH than SP spanning tree.

In other words, GridStix was predicted to be relatively resilient to node failure when configured to use WiFi and a fewest hop spanning tree, but at the cost of high power consumption.

Resilience and power consumption were two of GridStix's important non functional requirements. However, as shown in the experiments it is hard to optimize for both, meaning that one would have to be prioritized over the other. However, a feature of self-adaptive systems is that the extent to which any NFR must be satisfied (sufficiently satisfied) tends to be context-dependent, and this was the case with GridStix. Goal-based models, and specifically softgoals, support reasoning about tradeoff decisions that are aimed at achieving optimal goal satisfaction.

For the purposes of GridStix, expert environmental scientists had partitioned river behaviour into three distinct operating conditions (*domains*); *quiescent*, *high flow* and *flood*. Quiescence was predicted to be the most common domain over time and so, with the need for the nodes to retain enough power to react when the river state changed, energy efficiency was the priority. When in the flood and high flow domains, by contrast, resilience was prioritized to better tolerate any node loss that could impair the accuracy of GridStix's flood predictions. Thus, a particular GridStix configuration was specified for each domain, with (what was predicted to be) adaptation from one configuration to another specified to happen when the river was observed to change from one domain to another. These domain changes were based on sound knowledge and were therefore *foreseen*, meaning that we knew that the river's state would change and could specify the behaviour required of GridStix for each domain. Figure 5 shows the goal model for the flooding domain (which we call S3). The figure shows the claims "Bluetooth too risky for S3", "SP too risky for S3" and "Single node image processing not accurate enough for S3". Each claim records an assumption about a design-time choice of goal operationalization, made because of uncertainty about the relative performance of alternative operationalizations in the field. The tasks (goal operationalisation strategies) chosen are in white (i.e. WiFi, and FH). Note that for simplicity reasons the single-node and multi-node image processing shown in the figure is not part of the explanation. However, similar conclusions can be made if we take into account these operationalizations and their effect on the NFRs, therefore the *calculate flow rate* goal should be ignored in the figure.

The configurations that were specified at design-time for each domain were based on the performance of the alternative communication technologies and spanning tree configurations observed in the laboratory experiments described above. However, we were aware that the lab results might prove imperfect predictors of how GridStix performed in the field. The initial River Ribble deployment confirmed that the effects of radio signal absorption by the river banks, rain, trees, etc., had a significant affect on performance [21]. To make GridStix more tolerant of these effects, it was augmented with claims to monitor

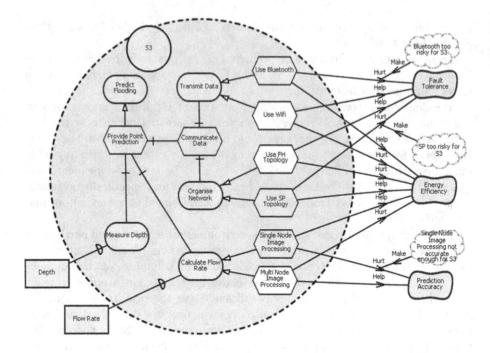

Fig. 5. Gridstix goal models for the flooding state of the river

the design-time assumptions, and to adapt to an alternative configuration if monitoring suggested that the alternative configuration could perform better. This was an important change because it meant that, in addition to the changes foreseen by knowledge of the different river domains, change as a consequence of operational experience was also *foreseeable*. When using claim monitoring, GridStix can decide by itself to adapt to a new configuration under some circumstances that were not predefined at design time. Thus, whereas GridStix's *adaptive* behaviour had been deterministic (even if its adequacy as a WSN had not been), its adaptive behaviour was now non-deterministic. Such non-deterministic behavior could cause "surprise" to an operator of the system, and therefore a self-explanation capability is appropriate.

A portion of the claim refinement models used by the GridStix *flood* and *high flow* domains is presented in Fig. 5. There is one top-level claim shown (in bold). This represents assumptions derived from the laboratory experiments that Bluetooth communication technology is too risky. In other words, the assumption is that if GridStix was configured to use Bluetooth, network resilience would likely be poor; implicitly poor*er* than if WiFi was used instead. The associated claim refinement model represents derivation of the means to sustain the claim and results in (using the labels in Fig. 6 as shorthand for the subclaims):

$$BT\_Too\_Risky \Leftrightarrow (A_0 \Leftrightarrow (A_1 \vee A_2)) \wedge (B_0 \Leftrightarrow (B_1 \Leftrightarrow (B_2 \vee \neg B_3)))$$

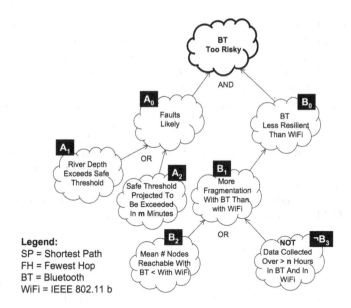

Fig. 6. GridStix claim refinement model justifying choice of WiFi for Inter-Node Communication

Thus, our root assumption, that using Bluetooth will lead to greater fragmentation than using WiFi (the *BT Too Risky* claim), will be disproved if any of the leaf (*monitorable*) subclaims is negated. In other words, Bluetooth is not likely to fragment the network if the river depth is below the safe threshold level or, at the current rate of change it will not exceed the safe level anytime soon. Similarly, Bluetooth is unlikely to lead to excessive fragmentation of the network if the rate of fragmentation when using Bluetooth is no higher than when using WiFi or, if there is data that contradicts this, there is too little data to make the contradiction statistically sound.

Because the River Ribble deployment of GridStix has been decommissioned, we used a simulator to observe the system's behaviour when experimenting with claim monitoring. The simulator has been developed using the collected data of the several months GridStix was deployed with the advantage that we can run experiments when needed. The simulator handles factors such as: power usage by batteries of nodes and according to whether the nodes were configured to use WiFi or Bluetooth, fewest hops or shortest path; whether the nodes were idling or performing computationally intensive tasks; and power replenishment from solar panels depending on time of day, amount of sunlight received or how cloudy the weather is, among others. Using a simulator constructed for GridStix, we ran an experiment to compare the longevity of the claim-augmented version of GridStix with the original. Longevity in this context means the length of time during which a sufficient number of nodes were connected to allow a meaningful result to be returned by the gateway. The simulator includes randomization to simulate jitter and packet loss. We complemented this with random node failures to simulate those actually observed. We ran the simulator with a profile of river behaviour over a fixed period comprising a sequence of flow rate and depth values that simulated the river in every mood from quiescent to flood. We varied a single variable; the amount of sunlight received by the nodes' solar panels, using percentage of cloud cover during daylight hours as a proxy. The experiment was run three times and the results averaged to account for the randomization elements.

The experiments suggest no significant benefit from claim augmentation when cloud cover is above approximately 40 %. Once cloud cover drops below 40 %, however, the augmented version has significantly greater longevity. For example, at 30 % cloud cover, instead of failing after approximately 180 h of operation, GridStix survives for approximately 250 h.

The increase in GridStix's longevity under some conditions appears to correlate with a particular self-adaptation being performed in these simulations. In those simulations where the claim-augmented version of the system outperformed the original version, GridStix substituted the use of WiFi for communication whilst the river was flooding, as originally specified, for the use of Bluetooth. The history of the monitoring data shows that over the defined minimum period of accumulating data, network fragmentation was no less during that period when using Bluetooth than when using WiFi. The effect of this on the claim refinement model (Fig. 6) in which the falsified monitorable claims $B_2 \vee \neg B_3$ ... became ... $\neg B_2 \wedge B_3$ and propagated up the hierarchy to falsify the top-level *BT Too Risky* claim that justified the original (design-time) choice of WiFi over Bluetooth. This in turn triggered the run-time re-evaluation of the goal model, revealing that the operationalization of the *Transmit Data* goal now favoured the use of Bluetooth rather than WIFi because Bluetooth's net impact on power consumption and resilience had become more +ve (positive) than that of WiFi. The goal model was thus changed to select Bluetooth as *Transmit Data*'s operationalization which in turn triggered the GridStix middleware to adopt a new component configuration, dynamically binding the Bluetooth component in place of the WiFi component (Fig. 7).

Discussion

Revisiting the challenges presented in the introduction of this paper we conclude that our approach:

1. Offers suitable monitoring capabilities for self-explanation through the use of claim monitoring. As for reasoning capability, our claim refinement models allow a change of configuration to be traced back to the monitored, and falsified, assumption that caused it.
2. Can offer self-explanation and discourse at the level of requirements. Self-adaptation (that maybe misunderstood by operators) can be explained in terms of goals, operationalisations and assumptions. This level of abstraction

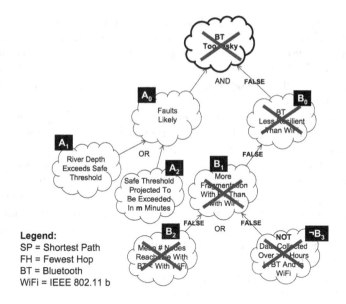

Fig. 7. Falsified claim propagation

is closer to natural language than architectural or code-level descriptions, offering more understandable explanations.

3. Provides the required link between the requirements and the architecture. The currently active configuration, at an architecture level, is linked to goal operationalisations, to the goals they achieve and their expected impact on system NFRs. (e.g. a proper link between the architectural use of BlueTooth and the "Communicate Data" goal it achieves, and the effect on system NFRs such as power consumption).

4. Finally, allows the system to be able to reproduce a trace history (i.e. a sequence of steps) of monitored events, assumption falsifications and resultant reconfigurations to explain the reasons behind a self-adaptation being carried out.

5   A Domain-Specific Language for Self-Explanation

An obvious goal for self-explanation research would be to offer natural language explanations, understandable by a system's users and other stakeholders. However, at present we feel that offering explanations in a simply-structured domain-specific language (DSL) is a more feasible goal. We propose a self-explanation DSL designed to convey explanations of self-adaptations in terms of requirements met, softgoals satisficed and claims invalidated, and supporting the identification of the run-time monitoring data that triggered a self-adaptation. Although explanations expressed in our DSL can be trivially transformed into text approaching natural language, knowledge and understanding of the system's requirements models is required to fully understand the explanation. Thus, our self-explanation DSL is targetted at support personel rather than end users. To extend our approach to come closer to the goal of offering user-understandable natural language self-explanations, some means of explaining the requirements models themselves would have to be devised.

As discussed in the previous section, self-adaptation in systems using our run-time requirements models can be either planned or unplanned. Planned self-adaptation fits the definition of foreseen adaptive behaviour in [18]. For unplanned self-adaptation, the acts of identifying an assumption, codifying it in a claim and devising a monitor for it imply a degree of foresight on the part of the developers that the system may need to tolerate conditions under which the assumption doesn't hold. However, the exact nature of the circumstances under which a claim (or some combination of claims) would be invalidated remain ambiguous, meaning that unplanned adaptation is most analogous to [18]'s foreseeable adaptive behaviour.

Planned self-adaptation takes place between a pair of pre-specified configurations, in response to a trigger conditions identified during the RE process. Planned self-adaptation requires the least detailed self-explanation, with the system needing to explain that the trigger conditions for a specific planned adaptation were met, and that the system has reconfigured itself to a specific, pre-specified configuration. This simpler self-explanation offers little more than standard logging output, and as such our DSL is designed to convey explanations of unplanned self-adaptation.

Unplanned self-adaptation takes place when a claim is invalidated, and the system's run-time requirements models are re-evaluated. Prior to an unplanned self-adaptation, the system may have adopted one of its pre-specified configurations, or another unplanned adaptation may have occurred, leaving the system in a configuration not foreseen at design time. A suitable self-explanation of an unplanned self adaptation needs to identify:

1. The configuration change that has been made.
2. The goal model change that resulted in the configuration change.
3. The softgoal(s) expected to be better satisfied in the new configuration.
4. The monitoring data that triggered the goal model change.

Explaining the details of the configuration change is a relatively simple matter, that could be achieved with more traditional logging or monitoring techniques. In terms of our run-time requirements models, the information required for the explanation is the task in the goal model whose selection to satisfy a parent goal is no longer supported by analysis of its contribution links and claims influencing them, and the task selected to replace it. Likewise, the monitoring data that triggered the goal model change is captured during the model transformation process, and could equally be captured through logging or monitoring.

The source of variability in our run-time requirements goal models is the invalidation of claims. Thus, the goal model change that resulted in an unplanned self-adaptation will be the invalidation of a specific claim, or combination of claims. In cases where claims are invalidated by propagation, the explanation should allow an indirectly invalidated claim's invalidation to be traced back to the original claim, whose invalidation triggered the propagation.

By default, the transformation to the run-time requirements models applied as a result of claim invalidation is an inversion of the claim's contribution link, by which it is connected to the model. Other transformations, such as the removal of claims or removing a malfunctioning goal operationalisation strategy, are supported. However, these other transformations either require further claims to be added to the model, or risk deadlock in the requirements model after claim invalidation, and we also prefer to invert claim contribution links where possible. For simplicity, we prefer to use a single root claim in the goal model to represent a specific area of uncertainty surrounding a variation point.

The softgoal(s) that are expected to be better satisfied as a result of the configuration change can be identified through model analysis. If the default (contribution inversion) transformation is used, and our model construction guideline of using a single claim per variation point has been followed, the promoted softgoal or softgoals are readily identifiable. Table 1 shows details, for each combination of the value of the contribution link connected to a claim (the "claim contribution") and the polarity of the contribution link to which the claim's contribution is attached (the "attached contribution"); which softgoal(s) can be expected to be better satisfied as a result of the self-adaptation.

In Table 1, the "Complementary" softgoals are defined as the softgoals connected to the task the attached contribution link belongs (and thus connects) to *with the same polarity.* Conversely, "Competing" softgoals are defined as those connected to the task the attached contribution link belongs with the *opposing polarity.*

An unplanned self-adaptation may also be triggered in response indicating that a claim that was previously invalidated does, in fact, appear to be valid once more. This would typically occur as a result of developers having missed some operating context encountered by the system only temporarily. We refer to this as claim revalidation. For this class of unplanned self-adaptation, the softgoal(s) expected to be better satisfied as a result of the configuration change are different to those in Table 1. Table 2 details the softgoals affected by claim revalidation, using the same columns and terminology as Table 1.

Table 1. Softgoals promoted by claim invalidation, for different model structures

Claim contribution	Attached contribution polarity	Softgoals affected
Make	Positive	Competing
Break	Positive	Complementary
Make	Negative	Complementary
Break	Negative	Competing

Table 2. Softgoals promoted by claim revalidation, for different model Structures

Claim contribution	Attached contribution polarity	Softgoals affected
Make	Positive	Complementary
Break	Positive	Competing
Make	Negative	Competing
Break	Negative	Complementary

Even if the explanation of an unplanned adaptation contains all of the information discussed so far, it may still prove difficult to understand the circumstances surrounding, and the reason for, an unplanned adaptation in cases where the system has previously performed another unplanned adaptation. In such circumstances, the configuration of the system prior to the adaptation under query may not be one of the configurations specified for the system at design time. Therefore, explanations of an unplanned self-adaptation must include the information of any previous unplanned self-adaptations performed by the system.

To summarise, our DSL is designed to convey self-explanations in terms of the satisficement of softgoals, reconfigurations, model transformations, claims invalidated, and monitoring events. Monitoring events are fired by monitors upon the collection and analysis of data indicating that a claim does not hold. Claims are invalidated as a result of monitoring data, and should the root claim in a claim refinement model (i.e. a claim upon which a decision rests) become invalidated then a model transformation is performed. A model transformation is a change to the run-time requirements model which may, if analysis of the modified run-time model indicates it is necessary, lead to a reconfiguration being performed by the system to better satisfice a softgoal. These concepts are all at the level of abstraction used in our requirements models, and thus this is the level of abstraction used by our explanations.

In Sect. 3, we proposed that a meaningful self explanation of a self-adaptive system's current behaviour needs to include:

1. Details of any change in priority, or the proposed degree of satisfaction of a system (soft)goal.
2. Details of the adaptation performed by the system.
3. The history of the adaptation, and the related events that triggered it.

In these terms, an explanation of a planned self-adaptation consists of details of the change in context identified by the system, the self-adaptation performed, and details of previous self-adaptations that have been performed. For unplanned self-adaptations, an explanation consists of details of the softgoals that are to be better satisfied by the adaptation, details of the configuration change performed, and details of the goal model changes that were made (claims invalidated) triggering the self-adaptation, along with the monitoring data that caused the goal model changes. Further history is also provided by including details of previous unplanned self-adaptations.

Our self-explanation DSL targets the more complex unplanned self-adaptations performed by a system, and presents the information discussed in tuples of:

{Softgoals Satisfied, Change Performed, Claim Invalidated, Cause}

A tuple is required for every change to the run-time requirements models performed by the running system. As a result, not all values in the tuple may be populated for every model transformation (e.g. the invalidation of an intermediate claim in a run-time Claim Refinement Model) as some information (e.g. the softgoals better satisficed by a self-adaptation) will not yet be available. The contents of the first three values in each tuple are as discussed so far in this section. Acceptable values for the "Cause" value in the tuple, however, vary depending on the type of change being made to the run-time requirements model. For a claim being invalidated directly by its own monitoring data, the cause is the monitoring data indicating the claim's invalidity. For claims invalidated by propagation, the cause identifies the claim invalidation propagated from, along with an indication of the Claim Refinement Model semantics dictating the claim be invalidated by propagation.

Analysis performed on a collection of tuples for an unplanned adaptation can yield a textual explanation, in terms of requirements, softgoals and claims, that approaches natural language. For the robot vacumm cleaner example discussed in Sect. 2, an explanation of the unplanned adaptation to clean the apartment when empty as opposed to cleaning at night would be explained as:

Adapted to "Clean When Empty" instead of "Clean at Night" to better satisfice "Avoid Tripping Hazard", due to the (invalidated) "No Tripping Hazard" claim. The "No Tripping Hazard" claim was invalidated because its supporting "No Foot Impact" claim was invalidated. The "No Foot Impact" claim was invalidated because monitoring data indicating its invalidity (FootShockEvent) was received.

Of course, our self-explanation DSL is tightly-coupled to our run-time requirements modelling approach, and to the use of claim invalidation as a source of run-time requirements model variability. However, our ability to generate self-explanations using our DSL from a running system, and to interpret them into a usable textual explanation serves to demonstrate the approach's feasibility. We also note that, at present, our method for generating output in the self-explanation DSL from a running system equipped with run-time requirements models using our run-time requirements models and reasoning tools depends on the (previously considered optional) one-claim-per-variation point modelling guideline being followed, and supports only one of three model transformations supported by the our run-time reasoner. We consider the use of more complex claim hierarchies, and particularly the "remove claim on invalidation" run-time model transformation with our self-explanation generator and DSL as interesting areas for future exploration.

6   Related Work

Although, to our knowledge, we are the first to discuss self-explanation in the context of self-adaptive systems; the desire for systems to produce output at a higher level to improve understanding is long-standing. For example, there has been significant research into Natural Language Generation by the Artificial Intelligence and Computational Linguistics communities.

In [22], Duggan and Bent present an algorithm, designed to infer the type of variables during compilation of programs written in implicitly typed languages such as ML or Haskell, where explicit variable type declarations are not used. The algorithm infers variable type by analysis of variable usage, annotating the program's syntax tree as it progresses. Inference is performed using a set of rules; for example a variable to which the addition operator is applied, with a right hand operand of 1, is an integer. A variable whose type is determined to be integer through this example rule would have the following explanation annotated to the program's syntax tree: +(x,1) gives x: int. Explanations can become considerably more complex when a variable's type is dependent on that of one or more other variables, however the base format remains the same. In this work, the explanation is used by the algorithm itself to allow explanation fragments previously generated to guide later type inferences, but the authors consider the approach potentially useful in providing debugging support.

Similarly, in [23], Van Baalen *et. al.* retrofit a domain specific code generator with explanatory capability for use at NASA. In this work, the explanation covers the relationship between a specification, domain theory and synthesised code. The explanation is relatively low-level, designed to allow developers to prove correctness, given NASA's obvious need for high-assurance software.

In [24], Huang and Fiedler discuss the PROVERB text planner, which verbalises mathematical (natural deduction) proofs. The planner uses a three-stage approach, with the first stage responsible for hierarchically decomposing complex proofs into a series of subproofs, the second stage identifies possible opportunities to "paraphrase" (or rather combine proof elements into larger, useful sentences) with the third stage actually generating the textual output. A more general overview of the state of the art in automated theorem provers, including discussion of the usefulness of their output, is offered in [25].

Although this work shows a research interest in providing high-level output to ease human understanding, our focus is not on programs providing natural language output, but in providing explanations of observed behaviour. Furthermore, the explanations offered by [22,23] are aimed at developers and mathematicians, respectively. The self-explanations we advocate are at a higher-level of abstraction, aimed at users and support personnel.

Debugging mechanisms, even those considered high-level [26,27], are focussed on data structures and code rather than on requirements, goals and operationalisation strategies. More closely-related work can be found in the field of requirements monitoring [4,5], from which we derive our claim monitoring. [28] proposes "awareness requirements", which are requirements that refer to the success or failure of other requirements. The authors state that awareness requirements may refer to goals, tasks, quality constraints and domain assumptions. Claim monitoring in our work is similar to domain assumption awareness requirements in [28], but their focus is on the mapping from requirements models to feedback loops, with no run-time representation of the awareness requirements.

The claim reasoning we use to demonstrate the utility of run-time requirements models in offering self-explanation is based on a combination of two previous streams of our work. In [13], we discuss the use of claims to highlight assumptions made during self-adaptive system specification, with a view to them being revisited in light of later requirement changes. In [10], we make the case for the run-time use of requirements models, with the ability to rectify deficiencies in requirements satisfaction using self-adaptation being a key motivator. Although we use reasoning of run-time claim refinement models to offer a limited form of self-explanation, the type of self-adaptive system claim reasoning proves most useful for are those with a limited number of potential goal operisational strategies, or where self-adaptation is being used to balance a set of conflicting non-functional requirements.

Approaches such as RELAX [17] and FLAGS [29] adopt fuzziness in requirements to allow self-adaptation to prioritise and optimise their satisfaction. In these approaches, a run-time requirements model could be used to record the (re)prioritisations that take place, and to allow explanations of adaptations in the context of which requirements were compromised and which favoured. Approaches such as [30], which use KAOS [31] goal models, could benefit from run-time analysis of obstacle models to offer self-explanation in terms of which obstacles have been detected in the operating environment, and which goal operationalisation strategies have been adopted to overcome them.

When tackling uncertainty, the ideas discussed in [32] are also related. As we do, the authors of [32] argue that uncertainty plays an important role in any software based system that needs adapt continuously to meet the goals. They argue that the focus of managing uncertain information should be on the rationale used to come to a decision. We emphasize the importance of being able to explain this rationale. In their case, the decision may be taken either during design or requirements (i.e. before execution). In our case, we go further because the self-adaptive system is able to make decisions at runtime as well. Finally, we believe our work is relevant to the implementation of dynamic traceability needed when dealing with self-adaptive systems where little work has yet been done. The authors of [33] discuss traceability in the presence of uncertainty. Similar to our work on claims, the authors of [33] propose to attach supplementary information to traceability links. This additional information describes the confidence and the rationale for its creation. The authors take into account the fact that the rationale that supports design decisions is often based on assumptions and beliefs. However, in contrast to our work, their work focuses on the case of software product lines and their evolution during software life cycle rather than on runtime adaptation

7   Conclusions and Future Work

This paper has argued that self-adaptive systems with the potential to behave in a manner not prescribed at design-time require self-explanation to allow emergent behaviour to be diagnosed, understood and explained. Self-explanation is important because it provides a means to increase confidence in, and resolve queries about, the behaviour of a self-adaptive system by its users. Self-explanation can also aid developers in understanding the behaviour of a self-adaptive system by tracing observed run-time behaviour (the *what*) to design-time assumptions, instrospect the strategy chosen (the *how*) and the extent to which they proved to be valid in operation (the *why*).

As already described in [12,16] we have developed an approach to creating self-adaptive systems capable of tailoring their behaviour to an operating environment not fully foreseen at design-time, using run-time requirements models. These systems are capable, indeed likely, to exhibit emergent behaviour. In this paper we show how self-explanation of such behaviour might be generated from the systems' adaptive reasoning machinery. The particular run-time requirements models used by our approach are in-memory representations of i* Strategic Rationale and NFR framework Claim Refinement Models, which are notably high-level in their nature. Our hypothesis is that these dynamic models, interpreted through the history of observed behaviour and adaptation events can provide a plausible means of explaining why the observed behaviour came about. This contrasts with the use of low-level reconfigurations and executed code paths used in standard debugging tools which are difficult to interpret in terms of systems' requirements, even for expert developers working on systems that don't have the added complexity of a self-adaptive capability.

We have demonstrated how a system equipped with our run-time requirements models for self-adaptation may also analyse these models to support self-explanation. We have introduced a simple self-explanation domain-specific language, and have shown that analysis by a running system of an explanation conveyed in our self-explanation DSL can yield a near natural language explanation in terms of requirements, claims and monitors.

There are several ways in which our approach can be improved. Currently, the claim reasoning and model transformation based adaptation mechanism discussed in this paper applies where goals are achieved by selecting from a finite number of goal operationalisation strategies defined a-priori but selected dynamically. Our approach is able to improve the flexibility of an executing system facing unforeseen situations, but the potential operationlization strategies, and the goals they achieve are defined and analysed at design time. Where new goal operationalisation strategies may themselves be emergent (e.g. through dynamic service discovery), further research is needed. This is one of the topics we are investigating in the FP7 CONNECT project[1]. Specifically, we are studying ways in which a new goal operationalisation strategy can be conceived at runtime.

---

[1] http://connect-forever.eu//

One of the challenges explored is that of updating goal models during execution to keep the required causal link between architecture and requirements.

Looking to more distant research prospects, two clear goals of self-explanation research are to provide natural language explanations that are within the capabilities of users to understand, and to allow users themselves to instruct the system to perform a self-adaptation by interacting with the system at this same level. The ability for a running system to use run-time requirements models to trace between high-level concepts such as goals and claims and lower-level system configurations, as demonstrated in this paper, indicates that this may be possible with future work.

References

1. Muir, B.M.: Trust in automation: Part I. theoretical issues in the study of trust and human intervention in automated systems. Ergonomics **37**(11), 1905–1922 (1994)
2. Gillieis, A.C., Hart, A.: Using kbs ideas in image processing - a case study in human computer interaction. In: Research and Development in Expert Systems V: Proceedings of Expert Systems '88, pp. 258–268 (1988)
3. Sawyer, P., Bencomo, N., Whittle, J., Letier, E., Finkelstein, A.: Requirements-aware systems: a research agenda for re for self-adaptive systems. In: IEEE International Conference on Requirements Engineering, pp. 95–103 (2010)
4. Robinson, W.: A requirements monitoring framework for enterprise systems. Requir. Eng. **11**(1), 17–41 (2005)
5. Fickas, S., Feather, M.: Requirements monitoring in dynamic environments. In: Second IEEE International Symposium on Requirements Engineering (RE'95) (1995)
6. Grace, P., Hughes, D., Porter, B., Blair, G., Coulson, G., Taiani, F.: Experiences with open overlays: a middleware approach to network heterogeneity. In: Submitted to Eurosys 2008, Glasgow, UK (2008)
7. Coulson, G., Blair, G., Grace, P., Joolia, A., Lee, K., Ueyama, J., Sivaharan, T.: A generic component model for building systems software. ACM Trans. Comput. Syst. **26**(1), 1–42 (2008)
8. Garlan, D., Cheng, S.W., Huang, A.C., Schmerl, B., Steenkiste, P.: Rainbow: Architecture-based self-adaptation with reusable infrastructure. IEEE Comput. **37**(10), 46–54 (2004)
9. Georgas, J.C., van der Hoek, A., Taylor, R.N.: Using architectural models at runtime to manage and visualize the adaptation process. In: Bencomo, N., Blair, G.S., France, R. (eds.) Models@run.time, Special Issue. IEEE Computer (2009)
10. Bencomo, N., Whittle, J., Sawyer, P., Finkelstein, A., Letier, E.: Requirements reflection: requirements as runtime entities. In: Proceedings of the 32nd ACM/IEEE International Conference on Software Engineering, ICSE '10, vol. 2, pp. 199–202. ACM, New York (2010)
11. Souza, V.E.S., Lapouchnian, A., Robinson, W.N., Mylopoulos, J.: Awareness requirements for adaptive systems. In: ICSE Symposium on Software Engineering for Adaptive and Self-Managing Systems, SEAMS 2011, Waikiki, Honolulu, HI, USA, 23–24 May 2011, pp. 60–69 (2011)
12. Welsh, K., Sawyer, P., Bencomo, N.: Towards requirements aware systems: Runtime resolution of design-time assumptions. In: Proceedings of the 26th IEEE/ACM International Conference on Automated Software Engineering (2011)

13. Welsh, K., Sawyer, P.: Requirements tracing to support change in dynamically adaptive systems. In: Glinz, M., Heymans, P. (eds.) REFSQ 2009. LNCS, vol. 5512, pp. 59–73. Springer, Heidelberg (2009)

14. Chung, L., Nixon, B.A., Yu, E., Mylopoulos, J.: Non-Functional Requirements in Software Engineering, vol. 5. Springer, Heidelberg (1999)

15. Yu, E.S.K.: Towards modeling and reasoning support for early-phase requirements engineering. In: RE '97: Proceedings of the 3rd IEEE International Symposium on Requirements Engineering (RE'97), Washington, DC, USA (1997)

16. Welsh, K., Sawyer, P.: Understanding the scope of uncertainty in dynamically adaptive systems. In: Wieringa, R., Persson, A. (eds.) REFSQ 2010. LNCS, vol. 6182, pp. 2–16. Springer, Heidelberg (2010)

17. Whittle, J., Sawyer, P., Bencomo, N., Cheng, B.H.C., Bruel, J.M.: Relax: a language to address uncertainty in self-adaptive systems requirement. Requir. Eng. 15(2), 177–196 (2010)

18. Andersson, J., de Lemos, R., Malek, S., Weyns, D.: Modeling dimensions of self-adaptive software systems. In: Cheng, B.H.C., de Lemos, R., Giese, H., Inverardi, P., Magee, J. (eds.) Self-Adaptive Systems. LNCS, vol. 5525, pp. 27–47. Springer, Heidelberg (2009)

19. Goldsby, H.J., Sawyer, P., Bencomo, N., Hughes, D., Cheng, B.H.: Goal-based modeling of dynamically adaptive system requirements. In: 15th Annual IEEE International Conference on the Engineering of Computer Based Systems (ECBS) (2008)

20. Lim, B.Y., Dey, A.K., Avrahami, D.: Why and why not explanations improve the intelligibility of context-aware intelligent systems. In: Proceedings of the 27th International Conference on Human Factors in Computing Systems, CHI '09, pp. 2119–2128. ACM, New York (2009)

21. Hughes, D., Greenwood, P., Coulson, G., Blair, G., Pappenberger, F., Smith, P., Beven, K.: Gridstix: Supporting flood prediction using embedded hardware and next generation grid middleware. In: 4th International Workshop on Mobile Distributed Computing (MDC'06), Niagara Falls, USA (2006)

22. Duggan, D., Bent, F.: Explaining type inference. Sci. Comput. Program. **27**, 37–83 (1995)

23. Van Baalen, J., Robinson, P., Lowry, M., Pressburger, T.: Explaining synthesized software. In: Proceedings of the 13th IEEE International Conference on Automated Software Engineering, ASE '98, pp. 240-249. IEEE Computer Society, Washington DC, USA (1998)

24. Huang, X., Huang, X., Fiedler, A., Fiedler, A.: Proof verbalization as an application of NLG. In: Proceedings of the 15th International Joint Conference on Artificial Intelligence (IJCAI), pp. 965–970. Morgan Kaufmann (1997)

25. Bundy, A.: Automated theorem provers: a practical tool for the working mathematician? Ann. Math. Artif. Intell. **61**, 3–14 (2011). doi:10.1007/s10472-011-9248-8

26. Golan, M., Hanson, D.R.: Duel - a very high-level debugging language. In: USENIX Winter, pp. 107–118 (1993)

27. Yang, J., Soffa, M.L., Selavo, L., Whitehouse, K.: Clairvoyant: a comprehensive source-level debugger for wireless sensor networks. In: Proceedings of the 5th International Conference on Embedded Networked Sensor Systems, SenSys '07, pp. 189–203. ACM, New York (2007)

28. Silva Souza, V.E., Lapouchnian, A., Robinson, W.N., Mylopoulos, J.: Awareness requirements for adaptive systems. Technical report, University of Trento (2010)

29. Baresi, L., Pasquale, L., Spoletini, P.: Fuzzy goals for requirements-driven adaptation. In: 2010 18th IEEE International Requirements Engineering Conference (RE), pp. 125–134 (2010)
30. Nakagawa, H., Ohsuga, A., Honiden, S.: Constructing self-adaptive systems using a kaos model. In: Second IEEE International Conference on Self-Adaptive and Self-Organizing Systems Workshops, 2008, SASOW 2008, pp. 132–137 (2008)
31. Dardenne, A., van Lamsweerde, A., Fickas, S.: Goal-directed requirements acquisition. Sci. Comput. Program. **20**, 3–50 (1993)
32. Lehman, M.M., Ramil, J.F.: Software evolution: background, theory, practice. Inf. Process. Lett. **88**, 33–44 (2003)
33. Anquetil, N., Grammel, B., Galvao, I., Noppen, J., Shakil, S., Arboleda, H., Rashid, A., Garcia, A.: Traceability for model driven, software product line engineering. In: ECMDA (2008)

A Higher-Order Agent Model with Contextual Planning Management for Ambient Systems

Ahmed-Chawki Chaouche[1,2]([⊠]), Amal El Fallah Seghrouchni[1],
Jean-Michel Ilié[1], and Djamel Eddine Saïdouni[2]

[1] LIP6 Laboratory, University of Pierre and Marie Curie,
4 Place Jussieu, 75005 Paris, France
{ahmed.chaouche,amal.elfallah,jean-michel.ilie}@lip6.fr
[2] MISC Laboratory, University Constantine 2,
Ali Mendjeli Campus, 25000 Constantine, Algeria
saidouni@misc-umc.org

Abstract. This paper presents a concrete software architecture dedicated to ambient intelligence (AmI) features and requirements. The proposed behavioral model, called Higher-order Agent (HoA) captures the evolution of the mental representation of the agent and the one of its plan simultaneously. Plan expressions are written and composed using a formal algebraic language, namely AgLOTOS, so that plans are built automatically and on the fly, as a system of concurrent processes. Based on a specific semantics, a guidance service is also proposed to assist the agent in its execution. Moreover due to the specific structure of AgLOTOS expressions, the update of sub-plans is realized automatically accordingly to the revising of intentions, hence maintaining the consistency of the agent.

Keywords: Ambient intelligence · BDI agent · Formal planning language and semantics · Dynamical plan revising · Planning consistency and guidance

1 Introduction

Ambient Intelligence (AmI) is the vision of ubiquitous electronic environment that is non-intrusive and proactive, when assisting people during various activities [1,2]. For the design of such complex systems, Multi-agent System (MAS) approaches offer interesting frameworks, since their agents are considered as intelligent, proactive and autonomous [3]. Thanks to their mental attitudes, the Belief-Desire-Intention (BDI) agents of [4] are able to use their Beliefs, Desires, and Intentions rationally, in order to select and execute a plan of actions.

The major problem for AmI agent consists in recognizing its environmental contexts, including its location and the discovery of other agents. Nevertheless, the HoA model proposed recently in [5], outlined how autonomous BDI agents can evolve and move within an ambient environment, based on a context-awareness. The major features and functionalities of AmI are taken into account,

© Springer-Verlag Berlin Heidelberg 2014
R. Kowalczyk and N.T. Nguyen (Eds.): TCCI XVI, LNCS 8780, pp. 146–169, 2014.
DOI: 10.1007/978-3-662-44871-7_6

in particular dynamic requirements such that: AmI systems can be open, thus agents can dynamically enter or leave the system.

The presented paper is inspired by the HoA model however an efficient planning management process is introduced in the architecture of the agent. We take profit from the fact that the plan of the agent can be derived from the current set of intentions of the agent. Our approach is based on a formal description language, namely *AgLOTOS*, allowing us to introduce modularity and concurrency aspects to compose sub-plans, viewed as processes. Unlike the formal description of [6], the AgLOTOS semantics overpasses the sequential execution of sub-plans. Rather, the concurrency of sub-plans is fully implemented and is only restrained with the purpose of solving possible inconsistency between intentions.

In this context, the planning process must be able to select and try one or more plans from some intention, and even deal with several intentions at the same time. Moreover, plans must be revised on the fly, in the sense that agents are dynamic entities which are changing the set of intentions and then plan, throughout their evolutions. As underlyied by several authors, this is considered an important notion in BDI agent conceptual frameworks [7–9].

The planning process we propose also aims at offering to each AmI agent, powerful predictive services, that can run on the fly. Like in other recent approaches which are dedicated to the planning and the validation of BDI MAS systems, e.g. [6,10], we focus on one agent rather than on the whole MAS, since this eases us to embed agent in whatever environment and to deal with the openness of AmI systems.

The original contributions of this paper are the following *(1)* a well-structure extension of the AgLOTOS language and its semantics, which allows automatic revisions of plans, *(2)* a concrete software architecture allowing the management of the possible actions failures and *(3)* automatic guidance service based on the representation of plans. The paper is organized as follows: Sect. 2 details the considered AmI features. This allows us to introduce our Agent model in Sect. 3, namely *Higher-order agent model (HoA)*, which captures the evolution of the agent in both its mental and planning states. In Sect. 4, the (structured) *AgLO-TOS* language is defined to build plans automatically from the set of intentions, based on a library of elementary plans. then, an efficient plan revising is provided in Sect. 4, which works accordingly to the revision of intentions. In Sect. 5, the semantics of AgLOTOS is enriched to automatically produce a *Contextual Planning System (CPS)*, in order to guide the execution of plans contextually. In Sect. 6, the description of our experiment project illustrates a concrete use of our approach. Section 7 presents the related works concerning the modeling and the specification of AmI agents and systems. The last section is our conclusion.

2   AmI Requirements

The AmI systems we consider are open space and dynamic systems. Their agents can reason and are assumed to be BDI agents thus have complex features as described in Fig. 1. An agent is assumed to be **autonomous** and **pervasive**, thus operate without the direct intervention of humans or other agents. It is **anticipative** and can process **rational** decisions, based on its own knowledge and beliefs.

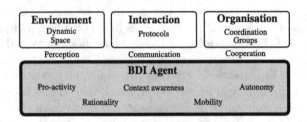

Fig. 1. Agent features for AmI systems

As a corollary of autonomy, an AmI agent is **context-aware**. We see the context of an agent as every environmental information perceived by the agent, in particular vicinity notions in a domain that considers space, time and social relationships. Thus, determining if a piece of information is relevant for an agent should be done based on its local context information.

To improve behavior and knowledge, an AmI agent can **communicate** and **cooperate** with its neighbors. Also, AmI agent can be **mobile** moving from one location to another one in a given space.

The BDI architecture is one of the major approaches to design pro-active agents in MAS [11]. Inspired from [4,12], the following functions describe the reasoning mechanism in the BDI agent, in order to produce a plan of actions. It is triggered by the perceived events (Evt) and often helped by a library of plans ($LibP$).

- $revs : 2^B \times Evt \rightarrow 2^B$ is the belief revision function applied when the agent receives a new event.
- $des : 2^B \times 2^D \times 2^I \rightarrow 2^D$ is the Desire update function that maintains consistency with the selected desires,
- $filter : 2^B \times 2^D \times 2^I \times LibP \rightarrow 2^I$ is the Intention function which yields the intentions the agent decides to pursue, among the possible options, taking into account new opportunities.
- $options : I \times LibP \rightarrow P$ is the function which associates a plan with each intention of the agent by using the $LibP$ library.
- $plan : 2^I \rightarrow P$ is a Plan function that processes an executable plan from some (filtered) Intention, knowing that any intention can be viewed as a partial plan.

3   The Higher-Order Agent Model

We are interested in modeling the evolution of the agent. Figure 2 highlights the agent architecture we consider in this paper, called *Higher-order Agent architecture (HoA)*. In contrast to the traditional BDI architecture, our approach enhances a clear separation in three processes:

- The *Context Process* is in charge of the context information of the agent. It is triggered by new perceptions of the environment and also by internal events informing about the executions of actions. At a low level, it is in charge of observing the realization of the executions of actions and plans of actions, in order to state whether they are successfully achieved or if a failure occurs.
- The *Mental Process* corresponds to the reasoning part of the agent. It is notified by the context process so that it can be aware of the important context changes and can provoke possible revisions of the beliefs (B), desires (D), and intentions (I) data.
- The *Planning Process* is called by the mental process. Helped by the LibP library, it mainly produces a plan of actions from the set of intentions, but also offers some services related to the management of plans.

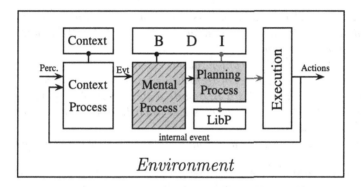

Fig. 2. Higher-order agent architecture

The behavioral model of the agent only relates on two main aspects of the agent: *(1)* the mental reasoning of the agent and *(2)* the evolution of the selected plan. Hence, we consider that the state of the agent, namely its *HoA configuration*, is a pair composed of a *BDI state* and a *Planning state*. As illustrated by Fig. 3, the occurrences of events may cause some changes of one of these parts or both.

The evolution of configurations is formally represented by the following *Higher-order Agent (HoA)* behavioral model. It is defined over an alphabet of events triggered by the actions being executed and by perception events, namely $Evt = Evt_{Act} \cup Evt_{Perc}$. Among the actions, message sendings are available, and message receivings are viewed as specific environmental perceptions. Moreover, mobility is handled as a specific action (*move*).

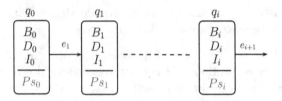

Fig. 3. The agent behavioral changes

Definition 1. *(The HoA model and HoA configuration)*

Consider any agent of the AmI system, then let $\mathcal{BDI}$ be the set of all the possible states that can be defined over the BDI structure. Moreover, let $\mathcal{P}$ be the set of all the possible corresponding plans, $\mathcal{PS}$ be the set of all the possible planning states evolving in some plan and let $LibP$ be a subset of $\mathcal{P}$ representing the library of plans. The HoA model of the agent is a transition system, represented by a tuple $\langle Q, q_0, \rightarrow, F_M, F_P, F_{PS} \rangle$, where:

- Q is the set of HoA configurations such that any configuration q is a tuple $q = (bdi, ps)$ where bdi and ps respectively represent the BDI and the planning states of the agent in q,
- $q_0 \subseteq Q$ is the initial HoA configuration, e.g. $q_0 = (bdi_0, ps_0)$,
- $\rightarrow \subseteq Q \times Evt \times Q$ is the set of transitions between configurations,
- $F_M : Q \longrightarrow \mathcal{BDI}$ associates a BDI state with each HoA configuration,
- $F_P : \mathcal{BDI} \times LibP \longrightarrow \mathcal{P}$ associates with each BDI state, an agent plan built from the $LibP$ library,
- $F_{PS} : Q \longrightarrow \mathcal{PS}$ associates a planning state with each HoA configuration.

In this paper, a BDI state is composed of three sets of propositions, representing the Beliefs, Desires and Intentions of the agent. The intentions are assumed to be partially ordered by associating a weight to each intention. This allows the mental process to organize its selected intentions, in order to solve possible conflicts. The Plan is directly derived from the intentions and is written as an AgLOTOS plan expression. Actually, the possible planning states are derived by using the semantics of AgLOTOS from the plan expression (see Sect. 4 to get more details).

A Simple AmI Example

Let Alice and Bob be two agents of an AmI Universitary system. Such a system is clearly open since agents can enter and leave. The fact that Bob is entering the system can be perceived by Alice in case she is already in. Since Alice is context-aware, she can take advantage of this information, together with other information like the fact she is able to communicate with Bob through the system.

Let $\Theta = \{\ell_1, \ell_2\}$ be two locations of the system where the agents behave. The proposed problem of Alice is that she cannot make the two following tasks in the same period of time: *(1)* to meet Bob in ℓ_1, and *(2)* to get her exam copies from ℓ_2. Clearly, the Alice's desires are inconsistent since Alice cannot be in two distinct locations simultaneously.

In the following sections, the *AgLOTOS* specification language will be used for formalizing this scenario.

4   Planning Formal Syntax and Semantics

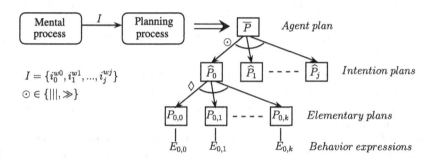

Fig. 4. Agent planning structure

Table 1. Synthetic presentation of the used notations

Notation	Description
q	HoA configuration
bdi	BDI state
ps	Planning state
E	AgLOTOS expression
P	Elementary plan
$\widehat{P}$	Intention plan
$\overline{P}$	Agent plan
(E, P)	Elementary plan configuration
$(E, \widehat{P})$	Intention plan configuration
$[\overline{P}]$	Agent plan configuration

In our approach, the planning language is well-structured as described in Fig. 4 and the associated notation in Table 1. Any plan of the agent, namely the *Agent plan* accords with the two following levels: *(1)* the agent plan is made of sub-plans called *Intentions plans*. Intentions plans correspond to the agent intentions and each one is dedicated to achieve the corresponding intention; *(2)* each intention plan can be an alternate of several sub-plans, called *Elementary plans*, each one being extracted from the LibP library. This allows one to consider different ways to achieve the corresponding intention (see Sect. 4.1). Further, we assume that the LibP library is indexed by the set of all the possible intentions that the agent can engage.

4.1 The Syntax of AgLOTOS Plans

Syntax of Elementary Plans. Elementary plans are written using the algebraic language *AgLOTOS* [13]. This language inherits from the LOTOS language [14] so offers different ways to express the concurrency of actions in plans. The building of an AgLOTOS expression refers to a finite set of observable actions. Further, let $\mathcal{O}$ be this set whose elements range over $a, b, \dots$ and let L be any subset of $\mathcal{O}$. Let $\mathcal{H} \subset \mathcal{O}$ be the set of the so-called AmI primitives which represent the mobility and communication:

- In AgLOTOS, actions are refined to make the AmI primitives observable: *(1)* an agent can perceive the enter and leave of another agent in the AmI system, *(2)* it can move between the AmI system locations, and *(3)* it can communicate with another agent in the system.
- An AgLOTOS expression refers to contextual information with respect to the (current) BDI state of the agent: *(1)* Θ is a finite set of space locations, *(2)* Λ is a set of agents with which it is possible to communicate, and *(3)* $\mathcal{M}$ is the set of possible messages to be sent and received.
- The agent mobility is expressed by the primitive $move(\ell)$ which is used to handle the agent move to some location ℓ ($\ell \in \Theta$). The syntax of the communication primitives is inspired from the semantics of π-calculus primitives, however with the consideration of a totally dynamic communication support, hence without specification of predefined channels: the expression $x!(\nu)$ specifies the emission to the agent x ($x \in \Lambda$) of some message ν ($\nu \in \mathcal{M}$), whereas, the expression $x?(\nu)$ means that ν is received from some agent x.

Let $Act = \mathcal{O} \cup \{\tau, \delta\}$, be now the considered set of actions, where $\tau \notin \mathcal{O}$ is the internal action and $\delta \notin \mathcal{O}$ is a particular observable action which features the successful termination of a plan.

The AgLOTOS language specifies a pair for each elementary plan composed of a name to identify it and an AgLOTOS expression to feature its behavior. Consider that elementary plan's names $\mathcal{P}$ are ranged over $P, Q, \dots$ and that the set of all possible behavior expressions is denoted $\mathcal{E}$, ranged over $E, F, \dots$. The AgLOTOS expressions are written by composing actions through LOTOS operators. The syntax of an AgLOTOS elementary plan P is defined inductively as follows:

$$
\begin{aligned}
P ::= \ &E & &\textit{Elementary plan} \\
E ::= \ &exit \mid stop & & \\
\mid \ &a; E \mid E \odot E & &(a \in \mathcal{O}) \\
\mid \ &hide\ L\ in\ E & & \\
\mathcal{H} ::= \mid \ &move(\ell) & &(\mathcal{H} \subset \mathcal{O}, \ell \in \Theta) \\
\mid \ &x!(\nu) \mid x?(\nu) & &(x \in \Lambda, \nu \in \mathcal{M}) \\
\odot = \ &\{\,[\,], |[L]|, |||, ||, \gg, [> \,\} & &
\end{aligned}
$$

The elementary expression *stop* specifies a plan behavior without possible evolution and *exit* represents the successful termination of some plan. In the syntax, the set $\odot$ represents the standard LOTOS operators: $E\,[\,]\,E$ specifies a non-deterministic choice, *hide L in E* a hiding of the actions of L that appear in E, $E \gg E$ a sequential composition and $E\,[>E$ the interruption of the left hand side part by the right one. The LOTOS parallel composition, denoted $E\,||[L]||\,E$, can model both synchronous composition, $E\,||\,E$ if $L = \mathcal{O}$, and asynchronous composition, $E\,|||\,E$ if $L = \emptyset$. In fact, the AgLOTOS language exhibits a rich expressivity such that the sequential executions of plans appears to be only a particular case.

Building of the Agent Plans from Intentions and Elementary Plans. The building of an agent plan requires the specific AgLOTOS operators:

- at the agent plan level, the parallel $|||$ and the sequential $\gg$ composition operators are used to build the agent plan, in respect to the intentions of the agent and the associated weights.
- the *alternate composition* operator, denoted $\Diamond$, allows to specify an alternate of elementary plans. In particular, an intention is satisfied iff at least one of the associated elementary plans is successfully terminated.

Let $\widehat{\mathcal{P}}$ be the set of names used to identify the possible intention plans: $\widehat{P} \in \widehat{\mathcal{P}}$ and let $\overline{\mathcal{P}}$ be the set of names qualifying the possible agent plans: $\overline{P} \in \overline{\mathcal{P}}$.

$$\widehat{P} ::= P \mid \widehat{P} \Diamond \widehat{P} \qquad\qquad \textit{Intention plan}$$
$$\overline{P} ::= \widehat{P} \mid \overline{P} \,|||\, \overline{P} \mid \overline{P} \gg \overline{P} \quad \textit{Agent plan}$$

With respect to the set I of intentions of the agent, the agent plan is formed in two steps: *(1)* by an extraction mechanism of elementary plans from the library, then *(2)* by using the composition functions called *options* and *plan*:

- *options* : $\mathcal{I} \to \widehat{\mathcal{P}}$, yields for any $i \in \mathcal{I}$, an intention plan of the form: $\widehat{P}_i = \Diamond_{P \in libp(i)}\, P$.
- *plan* : $2^I \to \overline{\mathcal{P}}$, creates the final agent plan $\overline{P}$ from the set of intentions I. Depending on how I is ordered, the intention plans yielded by the different mappings $\widehat{P}_i = options(i)$ for each $i \in I$, are composed by using the AgLOTOS composition operators $|||$ and $\gg$.

In our approach, the mental process can label the different elements of the set I of intentions by using a weight function *weight* : $I \longrightarrow \mathbb{N}$. This allows the planning process to schedule the corresponding intention plans yielded by the mapping *options*. The ones having the same weight are composed by using the concurrent parallel operator $|||$. In contrast, the intention plans corresponding to distinct weights are ordered by using the sequential operator $\gg$. For instance, let $I = \{i_0^1, i_1^2, i_2^1, i_3^0\}$ be the considered set of intentions, such that the superscript information denotes a weight value, and let $\widehat{P}_0, \widehat{P}_1, \widehat{P}_2, \widehat{P}_3$ be their corresponding intention plans, the constructed agent plan could be viewed (at a plan name level) as: $plan(I) = \widehat{P}_1 \gg (\widehat{P}_0 ||| \widehat{P}_2) \gg \widehat{P}_3$.

4.2   Semantics of AgLOTOS Plans

The AgLOTOS operational semantics is basically derived from the one of Basic LOTOS, which is able to capture the evolution of concurrent processes. A configuration (E, P) represents a process identified by P, such that its behavior expression is E. Table 2 recalls the Basic LOTOS semantics which formalizes how a process can evolve under the execution of actions. Here, it represents the operational semantics of *elementary plans*, viewed as processes. In particular, the last rule specifies how any (E, P) configuration is changed to (E', P) under any action a. Actually, $P := E$ denotes that the behavior expression E is assigned to P and $P \xrightarrow{a} E'$ represents the evolution of P from E to E'. Observe that for sake of simplicity, the notification that an action is launched is not represented in the semantics.

Table 2. Semantic rules of elementary plans

(Termination)	$\dfrac{}{exit \xrightarrow{\delta} stop}$
(Action prefix)	$\dfrac{a \in \mathcal{O}}{a;E \xrightarrow{a} E}$
(Choice)	$\dfrac{E \xrightarrow{a} E'}{F\,[]\,E \xrightarrow{a} E' \qquad E\,[]\,F \xrightarrow{a} E'}$
(Concurrency)	$\dfrac{E \xrightarrow{a} E' \qquad a \notin L \cup \{\delta\}}{E\,\|[L]\|\,F \xrightarrow{a} E'\,\|[L]\|\,F} \qquad\qquad \dfrac{E \xrightarrow{a} E' \qquad a \notin L \cup \{\delta\}}{F\,\|[L]\|\,E \xrightarrow{a} F\,\|[L]\|\,E'}$ $\dfrac{E \xrightarrow{a} E' \qquad F \xrightarrow{a} F' \qquad a \in L \cup \{\delta\}}{E\,\|[L]\|\,F \xrightarrow{a} E'\,\|[L]\|\,F'}$
(Hiding)	$\dfrac{E \xrightarrow{a} E' \qquad a \notin L}{hide\,L\,in\,E \xrightarrow{a} hide\,L\,in\,E'} \qquad\qquad \dfrac{E \xrightarrow{a} E' \qquad a \in L}{hide\,L\,in\,E \xrightarrow{\tau} hide\,L\,in\,E'}$
(Sequence)	$\dfrac{E \xrightarrow{a} E' \qquad a \neq \delta}{E \gg F \xrightarrow{a} E' \gg F} \qquad\qquad \dfrac{E \xrightarrow{\delta} E'}{E \gg F \xrightarrow{\tau} F}$
(Interruption)	$\dfrac{E \xrightarrow{a} E' \qquad a \neq \delta}{E\,[>\,F \xrightarrow{a} E'\,[>\,F} \qquad\qquad \dfrac{E \xrightarrow{\delta} E'}{E\,[>\,F \xrightarrow{\delta} E'}$ $\dfrac{F \xrightarrow{a} F'}{E\,[>\,F \xrightarrow{a} F'}$
(Relabeling)	$\dfrac{E \xrightarrow{a} E' \qquad a \notin \{a_1,...,a_n\}}{E[b_1/a_1,...,b_n/a_n] \xrightarrow{a} E'[b_1/a_1,...,b_n/a_n]}$ $\dfrac{E \xrightarrow{a} E' \qquad a = a_k \ (1 \leq k \leq n)}{E[b_1/a_1,...,b_n/a_n] \xrightarrow{b_k} E'[b_1/a_1,...,b_n/a_n]}$
(Plan definition)	$\dfrac{P := E \qquad E \xrightarrow{a} E'}{P \xrightarrow{a} E'}$

Definition 2 specifies how the expression of an *agent plan* is formed compositionally. Further, the behavior expression of the agent plan $\overline{P}$ is denoted $[\overline{P}]$ and is called the *agent plan configuration*. It is formed compositionally from the *intention plan configurations* of the agent, like $(E, \widehat{P})$ (see *rule 2*), themselves built from an alternate of *elementary plans configurations*, like (E_k, P_k) (see *rule 1*).

Definition 2. *Any agent plan configuration* $[\overline{P}]$ *has a generic representation defined by the following two rules:*

1.
$$\frac{\overline{P}::=\widehat{P} \qquad \widehat{P}::=\Diamond^{k=1..n}\, P_k \qquad P_k::=E_k}{[\overline{P}]::=(\Diamond^{k=1..n}\, E_k,\ \widehat{P})}$$

2.
$$\frac{\overline{P}::=\overline{P_1}\odot\overline{P_2} \qquad \odot\in\{|||,\gg\}}{[\overline{P}]::=[\overline{P_1}]\odot[\overline{P_2}]}$$

Table 3 shows the operational semantic rules defining the possible planning state changes for the agent. The rules apply from any HoA configuration $q = (bdi, ps)$, where the planning state ps is directly specified as an agent plan configuration, like $[\overline{P}]$. In each row of the table, there are three kinds of derivations: *(1)* the first rule shows the nominal case considering the execution of any action a $(a \in \mathcal{O} \cup \{\tau\})$, *(2)* and *(3)* the other two rules focus on the termination action of some intention plan, $\widehat{P}$. In the second one, the considered intention plan is successfully terminated whereas in the third one, the failure termination case is treated. With respect to any intention plan $\widehat{P}$, $\widehat{P}$ and $\neg\widehat{P}$ respectively represent the successful and failure termination cases of $\widehat{P}$. Hence, if $\mathcal{PS}$ is the set of all the possible planning states for the agent, then the transition relation between the planning states is a subset of $\mathcal{PS} \times Act \times (\widehat{\mathcal{P}} \cup \neg\widehat{\mathcal{P}}) \times \mathcal{PS}$. The transitions $(ps_1, a, \widehat{P}, ps_2)$ and $(ps_1, a, \neg\widehat{P}, ps_2)$ such that $\widehat{P} \in \widehat{\mathcal{P}}$, provoke an internal event informing the mental process of the termination of the intention plan $\widehat{P}$. For sake of clarity, the transition (ps_1, a, nil, ps_2) is simply denoted $ps_1 \xrightarrow{a} ps_2$, representing the execution of a non termination action a.

- The two first rows concern the derivations of the behavior expression of an intention plan $\widehat{P}$, under the execution of some action. The *Action rules* exhibit the simple case where E is an elementary expression of $\widehat{P}$, whereas the *Alternate rules* focus on the execution of an alternate of elementary expressions, like $\Diamond^{k=1..n} E_k$. The second rule captures the successful termination of $\widehat{P}$, under the execution of the action δ, while the third one captures the failure, in case the behavior expression of $\widehat{P}$ is equivalent to $fail$. In this paper, $fail$ represents the fact that the execution of some behavior expression E fails due to the dynamical context of the agent. In the first *Alternate rule*, the behavior expression $E = \Diamond^{k=1..n} E_k$ of an intention plan $\widehat{P}$, is refined by using the mapping *select*, the role of which is to select one of the elementary expression among the ones of E, e.g. $E_j = select(\widehat{P})$. The alternate operation is semantically defined by introducing a new semantic operator $\triangleright$, in order to take this selection into account: $E_j \triangleright (\Diamond E_{k\neq j}^{k=1..n} E_k)$. Observe that $E \triangleright F$, yields E if E is a success and F if E fails.
- In the two last rows, the sequential and parallel LOTOS operators are used to express the compositions of intention plan configurations, in a sequential or parallel way. Here again, the rules are refined to take possible failure cases into account.

Application to the Scenario. Let us take the scenario of Sect. 3 again. Table 4 separately represents a possible evolution of the HoA configurations for the agent

Table 3. Semantic rules of agent plan configurations

(Action)	$\dfrac{E \xrightarrow{a} E'}{(E,\widehat{P}) \xrightarrow{a} (E',\widehat{P})}$	$\dfrac{E \xrightarrow{\delta} stop}{(E,\widehat{P}) \xrightarrow[\widehat{P}]{\tau} (stop,\widehat{P})}$
	$\dfrac{E \equiv fail}{(E,\widehat{P}) \xrightarrow[\neg\widehat{P}]{\tau} (stop,\widehat{P})}$	
(Alternate)	$\dfrac{E_j \xrightarrow{a} E_j' \quad E_j = select(\widehat{P})}{(\Diamond^{k=1..n} E_k,\widehat{P}) \xrightarrow{a} (E_j' \rhd (\Diamond^{k=1..n}_{k\neq j} E_k),\widehat{P})}$	$\dfrac{E \xrightarrow{\delta} stop}{(E \rhd F,\widehat{P}) \xrightarrow[\widehat{P}]{\tau} (stop,\widehat{P})}$
	$\dfrac{E \equiv fail \quad F \xrightarrow{a} F'}{(E \rhd F,\widehat{P}) \xrightarrow{a} (F',\widehat{P})}$	
(Sequence)	$\dfrac{ps_1 \xrightarrow{a} ps_1'}{ps_1 \gg ps_2 \xrightarrow{a} ps_1' \gg ps_2}$	$\dfrac{ps_1 \xrightarrow[\widehat{P}]{\tau} ps_1'}{ps_1 \gg ps_2 \xrightarrow[\widehat{P}]{\tau} ps_1' \gg ps_2}$
	$\dfrac{ps_1 \xrightarrow[\neg\widehat{P}]{\tau} ps_1'}{ps_1 \gg ps_2 \xrightarrow[\neg\widehat{P}]{\tau} ps_1' \gg ps_2}$	
(Parallel)	$\dfrac{ps_1 \xrightarrow{a} ps_1'}{ps_1 \|\| ps_2 \xrightarrow{a} ps_1' \|\| ps_2 \quad ps_2 \|\| ps_1 \xrightarrow{a} ps_2 \|\| ps_1'}$	
	$\dfrac{ps_1 \xrightarrow[\widehat{P}]{\tau} ps_1'}{ps_1 \|\| ps_2 \xrightarrow[\widehat{P}]{\tau} ps_1' \|\| ps_2 \quad ps_2 \|\| ps_1 \xrightarrow[\widehat{P}]{\tau} ps_2 \|\| ps_1'}$	
	$\dfrac{ps_1 \xrightarrow[\neg\widehat{P}]{\tau} ps_1'}{ps_1 \|\| ps_2 \xrightarrow[\neg\widehat{P}]{\tau} ps_1' \|\| ps_2 \quad ps_2 \|\| ps_1 \xrightarrow[\neg\widehat{P}]{\tau} ps_2 \|\| ps_1'}$	

Alice and Bob, to solve the Alice's problem. In order to express the agent plan configurations, BDI propositions and plan of actions are simply expressed by using instantiated predicates, e.g. $get\_copies(\ell_2)$. Intention plans are composed from elementary plans which are viewed as concurrent processes, terminated by $exit$.

The initial configurations of Alice and Bob are respectively q_0^A and q_0^B, such that Alice is in ℓ_1 and has the mentioned two inconsistent desires, whereas Bob is in ℓ_2 and has expressed the desire to work with Alice. The current intention of Alice is only to meet with Bob. Here, BDI information is simply expressed by using intuitive predicate assertions.

The mental process can order the set of intentions to be considered. For instance, the intentions of Bob in q_1^B: $I_1 = \{meeting(Alice, \ell_1), getting\_copies(\ell_2)\}$ are ordered such that $weight(meeting(Alice, \ell_1)) < weight(getting\_copies(\ell_2))$. In the intention set I_1 of Bob, the corresponding agent plan configuration is: $[\overline{P_1}] = ((E_g, \widehat{P_g}) \gg (E_m, \widehat{P_m}))$, where $(E_g, \widehat{P_g})$ and $(E_m, \widehat{P_m})$ are the two intention plan configurations of Bob. The first one corresponds to the

Table 4. A state evolution for Alice and Bob

Alice's scenario
q_0^A $\begin{aligned} &B_0 = \{in(me, \ell_1), in(\boldsymbol{copies}, \ell_2)\} \\ &D_0 = \{meeting(\boldsymbol{Bob}, \ell_1), getting\_copies(\ell_2)\} \\ &I_0 = \{meeting(\boldsymbol{Bob}, \ell_1)\} \\ &[\overline{P_0}] = (meet(Bob); exit, \widehat{P_m}) \end{aligned}$
q_1^A $\begin{aligned} &B_1 = \{in(me, \ell_1), in(\boldsymbol{copies}, \ell_2), in(\boldsymbol{Bob}, \ell_2)\} \\ &D_1 = \{meeting(\boldsymbol{Bob}, \ell_1), asking(Bob, get\_copies(\ell_2))\} \\ &I_1 = \{meeting(\boldsymbol{Bob}, \ell_1), asking(Bob, get\_copies(\ell_2))\} \\ &[\overline{P_1}] = (meet(Bob); exit, \widehat{P_m}) \;\|\|\; (Bob!(get\_copies(\ell_2)); exit, \widehat{P_a}) \end{aligned}$

Bob's scenario
q_0^B $\begin{aligned} &B_0 = \{in(me, \ell_2)\} \\ &D_0 = \{waiting(\nu), meeting(\boldsymbol{Alice}, \ell_1)\} \\ &I_0 = \{waiting(\nu), meeting(\boldsymbol{Alice}, \ell_1)\} \\ &[\overline{P_0}] = (Alice?(\nu); exit, \widehat{P_w}) \;\|\|\; (move(\ell_1); meet(Alice); exit, \widehat{P_m}) \end{aligned}$
q_1^B $\begin{aligned} &B_1 = \{in(me, \ell_2), in(\boldsymbol{copies}, \ell_2)\} \\ &D_1 = \{meeting(\boldsymbol{Alice}, \ell_1), getting\_copies(\ell_2)\} \\ &I_1 = \{meeting(\boldsymbol{Alice}, \ell_1), getting\_copies(\ell_2)\} \\ &[\overline{P_1}] = (get\_copies(\ell_2); exit, \widehat{P_g}) \;\gg\; (move(\ell_1); meet(Alice); exit, \widehat{P_m}) \end{aligned}$

intention $getting\_copies(\ell_2)$ and the second to $meeting(Alice, \ell_1)$, such that $E_g = get\_copies(\ell_2); exit$ and $E_m = move(\ell_1); meet(Alice); exit$.

An example of execution derived from the initial planning state of Bob in q_1^B is the following, expressing that Bob fails to get the copies but this does not prevent him to move and perform the meeting with Alice:

$$((E_g, \widehat{P_g}) \gg (E_m, \widehat{P_m})) \xrightarrow[\neg P_g]{\tau} (E_m, \widehat{P_m}) \xrightarrow{move(\ell_1)} (E_m', \widehat{P_m}) \xrightarrow{meet} (E_m'', \widehat{P_m})$$

$$\xrightarrow[\widehat{P_m}]{\tau} (stop, \widehat{P_m}).$$

The reader may notice that the Alice and Bob agent plan configurations in q_1^A and q_1^B can change according to their revised intentions. Section 4.3 enriches the semantics with the plan revising service. This scenario will be taken again as an illustration.

4.3 Dynamical Plan Revising

In our model, the mental process drives the planning process such that adding or removing intentions possibly provoke the change of the agent plan. Of course, the mental process cannot ask for such a change on some plan is in progress, since this could imply that the agent could fall down in an incoherent state. In fact, the mental process must be informed by the planning process about the terminations of sub-plans, in particular intention plans, in order to act with consistency. At this point, we assume that the only dependencies within the intention set are due to the organization of the weighted intentions, required by the mental process. A rough approach would consist in waiting the whole plan termination, meaning that the planning process reaches the final planning state of the current agent plan, before taking the change of intentions into account. In this paper, we propose an improved method which consists in updating the agent plan as the revisings of intentions are required by the mental process. Such updates consist in adding new intention plans and removing some of the remaining intention plans in progress. We take profit from the compositional nature of AgLOTOS, that allows the planning process to manage the different intention plans distinctly. Recall that any planning state specifies different intention plan configurations corresponding to intentions to be achieved.

In some HoA configuration $q = (bdi, ps)$, $I(bdi)$ represents the intention set in this configuration. Let us consider some planning state $ps = [\overline{P}]$, such that $[\overline{P}] = \odot_{i \in 1..n} (E_i, \widehat{P_i})$ represents the remaining intention plan configurations to execute, where each $\widehat{P_i}$ corresponds to the plan of the intention i, E_i is its associated AgLOTOS expression and $\odot \in \{|||, \gg\}$.

The updatings of the intention set and of the corresponding agent plan rely on the following principles:

- according to the semantics, every termination of intention plan produces an internal event which changes the BDI state, in particular the updating of the intention set.
- it is easy to build some mappings which relates every intention $i \in I$ to the corresponding pair $(E, \widehat{P})$ and vice versa: (1) $remain : \overline{P} \times I \to \mathcal{E} \times \widehat{P}$ maps each intention $i \in I$ to the corresponding pair $(E, \widehat{P})$ of $[\overline{P}]$; (2) $index : \widehat{P} \to I$ maps each $\widehat{P}$ to the corresponding intention $i \in I$. It is worth noting that from $weight(index(\widehat{P_i}))$ such that $i \in I$, one yields the weight of the intention i.

The add and remove of updating operations are formalized by the following two mappings: $add, remove : 2^{\mathcal{E} \times \widehat{P}} \times \mathcal{I} \to [\overline{P}]$, be the mapping which builds an agent plan configuration $[\overline{P}]$ from a set of intention plan configurations.

The *Adding* of a new intention k, assuming its intention plan configuration is $(E_k, \widehat{P_k})$, means:

- adding k in I and updating the weight mapping to take k into account, then
- building a new agent plan configuration, from the set of remaining intention plan configurations $\cup_{i \in I} remain(\overline{P}, i)$ and their respective weights $weight(i)$.

Formally, let ps be the current planning state of the agent and k be the intention to be added, the resulting planning state after the revising is defined by: $(add(\cup_{i \in I} remain(ps, i), k))$.

The explicit *Removing* of a (non-terminated) intention k from I, means that the corresponding $(E_k, \widehat{P_k}) = remain(\overline{P}, k)$ must be removed from $\overline{P}$. As for the adding function the resulting planning state after removing the intention k is: $(remove(\cup_{i \in I} remain(ps, i), k))$.

Application to the Scenario. Consider the example of Table 4 again, the changes of configurations for Alice and Bob (taken separately) are due to the respective perceptions of Alice and Bob and the fact they are anticipative. Actually, after having perceived that Bob is in ℓ_2 ($e_1 = perc(in(Bob, \ell_2))$), meaning in the same location as the exam copies, Alice enriches her beliefs, desires and intentions, aiming communicating with Bob and asking for his help to bring her the copies. Consequently, she evolves to the new configuration q_1^A, where the generated plan suggests that Alice sends the message $Bob!(get\_copies(\ell_2))$.

Notice that Bob is able to receive any message from Alice, which is denoted $Alice?(\nu)$ in q_0^B. The reception of the message sent by Bob triggers an event at the Bob's mental process level. Since here Bob is accepting bringing the copies to Alice, he expands his beliefs ($in(copies, \ell_2)$) and also considers a new intention $getting\_copies(\ell_2)$ to take into account, in fact consistent with the previous ones. Consequently, the HoA configuration of Bob is changing to q_1^B i.e. $I_1 = I_0 \cup \{getting\_copies(\ell_2)\}$, and the plan $[\overline{P_0}]$ of Bob is updated by using the add mapping: $[\overline{P_1}] = add(\cup_{i \in I_0} remain(\overline{P_0}, i), getting\_copies(\ell_2))$, in order to satisfy all of his desires, getting first the copies then going to meet Alice.

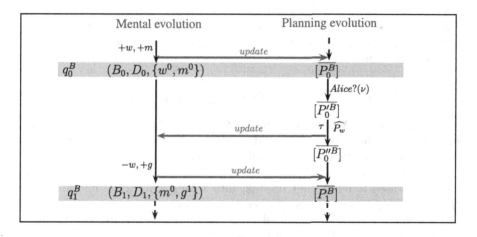

Fig. 5. HoA evolution of the agent Bob

The sequence diagram of Fig. 5 focuses on the behavior of Bob, in order to highlight the mutual updates required by the mental process and the planning process, to synchronize the content of the intention set and the specification of the agent plan. For sake of simplicity, notice that:

- w and m respectively represent the two initial intentions of Bob in I_0, that are $waiting(\nu)$ and $meeting(Alice, \ell_1)$,
- g represents the intention $getting\_copies(\ell_2)$

Moreover, the signs $+$ and $-$ respectively mean adding the intention and removing it.

The bold horizontal arrows show the updates of this part of the scenario. First, the agent plan is built from the fact that w and m are added to the intentions set (with the same weight). This results in two concurrent intention plans, $[\overline{P_0^B}] = (E_w, \widehat{P_w}) \| \| (E_m, \widehat{P_m})$. Due to the reception of the Alice's message, this yields to the execution of the reception action, followed by the exit action (τ) mentioning the termination of the intention plan $\widehat{P_w}$. This yields the agent plan $[\overline{P_0''^B}] = (E_m, \widehat{P_m})$. Since the mental process is informed of both actions (by internal events), it finally decides to update its set of intentions, with $-w, +g$ (m has a lower weight than g). As a consequence, the planning process is triggered to revise its agent plan in accordance, so $[\overline{P_1^B}] = (E_g, \widehat{P_g}) \gg (E_m, \widehat{P_m})$.

5   Contextual Planning Services

We now augment the planning process by two new services, exploiting the contextual information of the agent. Our aim is to provide a guidance service for the mental process and a model checking approach to analyze some temporal properties over the agent plan. Both services are based on the building of a specific transition system called *Contextual Planning Systems* (*CPS*), that can represent the different execution traces the agent could perform, from a given HoA configuration, in the best case, assuming the information context of the HoA configuration hold.

5.1   Building of a Contextual Planning System

Let consider any HoA configuration, the *CPS state* of the agent is now defined contextually to this configuration, taking into account the agent location and a termination information about the different intention plans.

Let q be any HoA configuration of the agent, and $I(q)$ its intention set in q, the rules used to build the CPS, take into account contextual information of three kinds which are: *(1)* the reached location in a CPS state, *(2)* the set of intention plans that are terminated when reaching a CPS state, and *(3)* more globally, the set $\Lambda(q)$ of neighbors currently known by the agent.

Definition 3. *A CPS state is a tuple* (ps, ℓ, T)*, where ps is the current planning state of the agent which represents its agent plan configuration* $[\overline{P}]$*, ℓ corresponds to the location within which the agent is placed, and T is the subset of intention plans which are terminated.*

Table 5 brings out the operational semantic rules, defining so the possible ways of CPS state changes. These rules are applied from an initial CPS state and $([\overline{P}], \ell, \emptyset)$ means that the agent is initially considered at location ℓ, and its plan configuration is $[\overline{P}]$.

The *Action rules* are used to derive the execution of an action with respect to an intention plan. The left hand side rule exhibits the case of a regular action, whereas the right hand side one shows the termination case of an intention plan, wherein the terminated intention plan is added to T. Also, the *Mobility rules* capture the change of location from ℓ to ℓ' and in the *Communication rules*,

Table 5. Semantic rules of CPS state changes

(Action)	$\dfrac{ps \xrightarrow{a} ps' \quad a \in \mathcal{O} \cup \{\tau\}}{(ps,\ell,T) \xrightarrow{a} (ps',\ell,T)}$	$\dfrac{ps \xrightarrow{\tau} ps'}{\widehat{P}} \quad \dfrac{}{(ps,\ell,T) \xrightarrow{\tau} (ps',\ell,T \cup \{\widehat{P}\})}$
(Mobility)	$\dfrac{ps \xrightarrow{move(\ell')} ps' \quad \ell \neq \ell'}{(ps,\ell,T) \xrightarrow{move(\ell')} (ps',\ell',T)}$	$\dfrac{ps \xrightarrow{move(\ell)} ps'}{(ps,\ell,T) \xrightarrow{\tau} (ps',\ell,T)}$
(Communication)	$\dfrac{ps \xrightarrow{x!(\nu)} ps' \quad x \in \Lambda}{(ps,\ell,T) \xrightarrow{x!(\nu)} (ps,\ell,T)}$	$\dfrac{ps \xrightarrow{x?(\nu)} ps' \quad x \in \Lambda}{(ps,\ell,T) \xrightarrow{x?(\nu)} (ps',\ell,T)}$

the action send $x!(\nu)$ (resp. receive $x?(\nu)$) is constrained by the visibility of the agent x in its neighborhood.

Observe that due to the fact we consider a predictive approach in this section, only successful executions are taken into account, thus abstracting that a plan may fail. Moreover, the semantics of the alternate operator is reduced to a simple non-deterministic choice of LOTOS: $\lozenge^{k=1..n} E_k \equiv [\,]^{k=1..n} E_k$ in order to possibly try every elementary plan to achieve the corresponding intention.

Definition 4. *Let $q = (bdi, ps)$ be any HoA configuration of the agent, the Contextual Planning System of q, denoted $CPS(q)$, is a labeled kripke structure $\langle S, s_0, Tr, \mathcal{L}, \mathcal{T} \rangle$ where:*

- *S is the set of CPS states,*
- *$s_0 = (ps, \ell, \emptyset) \in S$ is the initial CPS state, such that $ps = [\overline{P}]$ represents the current planning state of the agent in q and ℓ its current location,*
- *$Tr \subseteq S \times Act \times S$ is the set of transitions. The transitions are denoted $s \xrightarrow{a} s'$ such that $s, s' \in S$ and $a \in \mathcal{O} \cup \{\tau\}$,*
- *$\mathcal{L} : S \to \Theta$ is the location labeling function,*
- *$\mathcal{T} : S \to 2^{\widehat{P}}$ is the termination labeling function which captures the terminated intention plans in some CPS state.*

Moreover, in a CPS, the transitions from a CPS state only represent actions that are realizable. In this paper, actions are modeled by instantiated predicates and their execution in CPS state is submitted to pre-conditions to be satisfied with respect to the contextual information known in that state, e.g. $pre(x!(\nu)) = pre(x?(\nu)) = x \in \Lambda$.

Remark 1. (Realizable actions) Let $pre(a)$ be the pre-condition of the action a. A transition $(s \xrightarrow{a} s') \in Tr$ is realizable in the state s of $CPS(q)$ iff $pre(a) \subseteq \mathcal{L}(s) \cup \Lambda(q)$.

Consider Table 4 where $[\overline{P_1}]$ is the agent plan configuration considered for Bob in the HoA configuration q_1^B. The initial CPS state s_0 in q_1^B is written: $s_0 = ([\overline{P_1}], \ell_2, \emptyset)$ where $[\overline{P_1}] = ((E_g, \widehat{P_g}) \gg (E_m, \widehat{P_m}))$. An example of trace of $CPS(q_1^B)$ derived from s_0 is: $((E_g, \widehat{P_g}) \gg (E_m, \widehat{P_m}), \ell_2, \emptyset) \xrightarrow{getc} ((E'_g, \widehat{P_g}) \gg (E_m, \widehat{P_m}), \ell_2, \emptyset) \xrightarrow{\tau} ((E_m, \widehat{P_m}), \ell_2, \{\widehat{P_g}\}) \xrightarrow{move(\ell_1)} ((E'_m, \widehat{P_m}), \ell_1, \{\widehat{P_g}\}) \xrightarrow{meet} ((E''_m, \widehat{P_m}), \ell_1, \{\widehat{P_g}\}) \xrightarrow{\tau} (stop, \ell_1, \{\widehat{P_g}, \widehat{P_m}\}).$

5.2   Planning Guidance

In a CPS, the mapping $\mathcal{T}(s)$ captures the termination of intention plans in the CPS state s, hence the satisfaction of the corresponding intentions. In order to guide the agent, the planning process can select an execution trace through the plan, which maximizes the number of intention terminations.

Definition 5. *(Maximum trace) Let end* $: \Sigma \longrightarrow 2^{\widehat{P}}$ *specify the set* $end(\sigma)$ *of termination actions that occur in a trace* σ*, then, the trace* σ *is said to be maximum iff there is no trace* σ' *in* $CPS(q)$ *such that* $|end(\sigma')| > |end(\sigma)|$.

As a corollary, a similar technique can be used to check the consistency of the agent intentions. With respect to any HoA configuration q, the comparison between one *maximum trace* σ in $CPS(q)$ and $|end(\sigma)|$ allows one to exhibit two extreme cases:

– if $|end(\sigma)| = |I(q)|$, we conclude that all the intentions of $I(q)$ are consistent,
– if $|end(\sigma)| = 0$, there is no satisfied intention, so the agent plan $\overline{P}$ of q is unappropriated with respect to the set of agent intentions.

We reconsider the scenario of Table 4 to achieve the intentions of Bob in a parallel way: $[\overline{P_1}] = ((E_g, \widehat{P_g})|||(E_m, \widehat{P_m}))$ is the agent plan configuration considered for Bob. The Contextual Planning System $CPS(q_1^B)$ is highlighted in

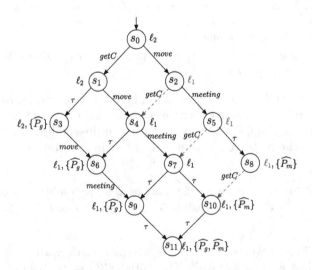

Fig. 6. $CPS(q_1^B)$ corresponding to the plan $\overline{P_1}$ of Bob

Fig. 6. It is built from the initial CPS state $([\overline{P_1}], \ell_2, \emptyset)$ in q_1^B. In the figure, the dashed edges represent the unrealized transitions from the states $s \in \{s_2, s_5, s_8\}$, because $pre(getC) = \{\ell_2\} \not\subseteq \mathcal{L}(s) \cup \Lambda(q)$.

Model Checking of a Plan Properties

The fact that $CPS(q)$ is a kripke structure, allows one to process temporal logic verification, such as LTL or CTL. Due to the contextual labeling of states in $CPS(q)$, the properties to be checked concern the mobility, the communication and the terminations of intention plans. For instance, with respect to the $CPS(q_1^B)$ of Fig. 6, the CTL-formula $\mathrm{AF}(\ell_1)$ is checked **true**. This property means that the agent eventually reaches or crosses the location ℓ_1 on all the possible traces featured in the CPS. The use of model checking techniques is rather large since the possible properties cover both safety and liveness considerations. In fact, we view the CPS-based model checking service as a way to reinforce the planning guidance.

6 Experimentation: The Smart-Campus Project

We experiment our agent-based approach in a distributed system project called *Smart-Campus*. The aim is to assist the users of a complex Universitary campus in their activities. Concretely, we equip a float of (Android) smart-devices by the smart-campus application. In this application, the software architecture is composed of an HoA agent and a specific graphical user interface (GUI) to interact with the user to be assisted. Basic services are currently implemented over the HoA architecture, based on physical localization and (a)synchronous communication mechanisms. They are supported by the smart-device API facilities, in particular the Wireless Local Area Network (WiFi) API. As an example, the navigation service takes profit from the underlyied localization service to determine on the fly, the position and the move of the assisted user. Moreover, the contextual guidance service allows the agent to assist the user in realizing his desires in proposing different alternatives.

At the level of the Smart-Device (SD), the campus is concretized by the starting service which automatically connects the SD to the "CAMPUS" network, through one of the possible WiFi Access Points (AP). As illustrated in Fig. 7, the SD can automatically access to the server *"SC Directory"*, dedicated to the smart-campus. This server is viewed as a middleware which maintains the persistence of contextual information like the discovery and the locations of other SD and objects concerning the campus. The starting service is also dedicated to declare the public information of the user to the server, in particular its location. One of the specificity of this project is that the agent embedded in the SD remains autonomous when the SC directory cannot be reached or when the user is exiting the campus. It can continue assisting the user, due to the context information and persistent data previously stored in the SD, that can be pervasively updated with the help of other neighbor agents.

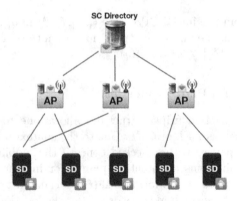

Fig. 7. Smart-campus architecture

Observe that the localization service must work over the campus ground as well as the different stairs of the buildings. The best localization indoor technique is currently a research in progress e.g. [15]. Currently, we use different WiFi access points within the campus to compute the geographical locations, since this works in both indoor and outdoor locations. Anyway, the localization process requires a tune calibration phase to store specific information in the SC directory, concerning a set of physical reference points that must be selected over the campus, as mentioned in the finger printing approach, e.g. in [16].

In our case, information includes the physical location of the reference point (GPS), its symbolic name (place/room/corridor) and above all the perceived signal attenuation (RSSI[1]) from that location, in respect to the different WiFi access points. The localization service on the SD can then compare its proper perceptions of the WiFi attenuation in respect to the same references stored in the server, so that to deduce an approximation of its position through statistical computations and trilateration concepts.

GUI is also an important issue in our application since the user is definitively a mental process and also have physical capabilities expressed by all its senses. In particular, the move action finally results in some notification/proposal to the user which can walk in the good direction. Actually, all the actions that the HoA agent cannot perform can be delegated either to other agents or to some connected user, accordingly to the required capacities. As a specific GUI, graphical maps are modern and useful interfaces for the users. Our application is able to manage the maps of the campus, over which additional layers are used to render maps interactive and to show different locations and paths.

7   Related Work

Since the last decade, several MAS projects were proposed embedding AmI systems. In conjunction with them, an important issue was to understand how AmI

---

[1] RSSI: Received Signal Strength Indication.

agents can interact and reason within a dynamic environment. One important topic was thus to model context-awareness within agent. Among the different existing works, different management of contexts were proposed: In [17], a hierarchical space system is considered which allows to directly specify location and some context elements for each agent moving in the space. Moreover, a list of interesting AmI system projects is brought out in the paper. In [17], the system view is abstracted but the agent context is modeled as a dynamic structure over which coordination activities are recognized by a pattern matching technique.

Among the different existing works, different management of contexts were proposed, and were dedicated to the description of AmI applications and scenarios, e.g. in [1–3]. These works concentrate on engineering issues of AmI systems while taking into account sophisticated contexts. In our formal approach, the context we deal with only refers to location and neighboring information, however, it could be extended with additional managements and techniques.

The fact to consider MAS as distributed systems composed of different communicating entities is not new, however the first proposals either do not include any built-in capacity for "lookahead" type of planning or they do it only at the implementation level without any precise defined semantics. Since 2006, BDI agent-centric approaches emerge to cope with the dynamicity of the agent environment. In [6], a hierarchical model (HTN) is proposed to better control the scheduling of plans in BDI agents, following an alternating goal-plan oriented strategy. Later on the same bases, Sardina et al. [18] showed how to specify and test learning approaches in some particular cases. In [8], they proposed a formal procedure to manage and update the intention sets according to the agent BDI attitudes and the occurrences of new events.

Our proposed model is also agent-centric and accords with the principle of tightly controlling plan from the BDI mental attitudes. In contrast to the former works, the realizations of different intentions can be simultaneously considered and concurrently executed, in a higher level planning model. Conflicts are assumed to be solved by the mental process of the agent, however with the help of intention priorities. It is worth noticing that the conflicts which are caused by the contextual information are taken into account when building a CPS from the intention set. Dealing with action dependencies to restrain the agent activity have already largely been studied in the literature. In particular a GraphPlan planner can efficiently produce a global plan as a flow of actions, that corresponds to the subset of the desires that could be executed concurrently [10]. However, GraphPlan only deals with some of the possible schedulings between actions, since it follows a global time step approach. In contrast, our approach takes all the possible cases into account.

The on the fly revision of plans throughout the evolution of the agent is in fact a recent research topic in MAS approaches, e.g. [19]. In [8], the authors propose to revise the agent plans from the modifications of the agent's goals, in particular from the fact that some plans corresponding to intentions can be conflictual. However, the proposed solution comes from the fact that actions are processed atomically, so without concurrency, and in a tight alternate of goals and basic actions, expressed in a hierarchical structure. The well-structured semantics of AgLOTOS, especially with the intermediary notions of intention plans allows us to smoothly develop the planning update service from some changes in the set of intentions. It is not related to the concurrency of intention plans and has the capacity of continuing the execution of intention plans which are not related to the change of some intention.

In the literature there are a number of (BDI) agent programming languages [20], highlighting different aspects or modules developed in agent softwares like goal, planning, organization, reinforcement learning, ..., e.g. S-CLAIM [21], Jason [22], 2APL [23] and JIAC [24]. Most of them lack from formal description, but, the work of [6] has extended some existing formal algebraic specification models dedicated to distributed systems, in order to specify the concurrency of actions in plans. The proposed language integrates BDI ingredients within plans in a unified and interactive way. Nevertheless, our formal algebraic language based on LOTOS, appears to be more expressive in its capacity to express plans as concurrent processes as well as concurrent actions. It is also operational by the way to handle action and plan failures in the AgLOTOS language and the HoA architecture. Lastly, pay attention that in our approach, a clear separation exists between the mental and planning levels. Actually, our planning process behaves like a service that could be embedded in existing BDI architectures.

The verification task standardly applied to MAS is mainly driven by a global vision of the system, e.g. in [25]. In [26], the reuse of some program checking techniques are proposed, based on a BDI representation of the system state space. Moreover, in order to cope with the well-known combinatorial explosion of states, abstraction/reduction techniques are applied over the BDI states. One could think to introduce similar techniques within AmI agents, however, the high-level dynamics that usually features an AmI environment could introduce too much states to consider, even after reduction. Moreover, it remains open to apply these techniques according to the changes of contexts. In our paper, the proposed techniques, guidance and plan model checking, capture the AmI context but remains agent-centric. This avoids considering the system combinatorial explosion problem while allowing to take into account the AmI system dynamicity.

Petri net-based models were proposed to represent some MAS paradigms. Basically, such models can easily capture resource notions and concurrency execution of actions and dynamic processes, e.g. [27,28]. With some typing mechanisms (coloration), MAS distributed entities can be designed and their way to communicate between them, e.g. [29]. Thus, agent interactions can be handled at different operational levels, like agent plans [30]. As interesting works, the Petri Net modeling were useful to study the global coherencies of plans or agent interactions to recognize some FIPA protocols [31]. Nevertheless, many problems remain open in order to capture a whole AmI system requirements such as the system openness or the dynamics inherent to the change of agent contexts. In fact, Petri Nets remain low level representations with respect to the concepts used to build the mental attitudes of the agent. In contrast, AgLOTOS brings out abstraction capacity and high level composition means whereas the proposed HoA model allows to correlate both mental and planning levels.

8   Conclusion

The Higher-order agent model (HoA) formally represents a BDI-AmI open system where agents can reason, communicate and move. Agent dynamicity and context-awareness are handled due to the fact that a BDI agent can change its BDI state adequately to the perceptions of new events. The proposed AgLOTOS process-based algebra appears to be a powerful and intuitive way to express an agent plan as a set of concurrent processes. The presented scenario shows how the AgLOTOS language is rich due to modularity concepts and concurrency operators.

We demonstrate that the proposed model is operational. Actually, plans are automatically built from the set of intentions of the agent and a library of elementary plans already expressed in AgLOTOS. In contrast to existing works, the planning process can execute the different intention sub-plans concurrently. Moreover, the planning structure appearing in the AgLOTOS expressions allows to automatize the plan revisions on the fly, accordingly to the updates of intentions. Hence, an AmI agent can be viewed as having only one plan, updated all along the evolution of the agent behavior.

On another point, the CPS structure appears to be an interesting state-transition structure to select optimal execution from the contextual situation of the agent. In this paper, the proposed guidance service based on the CPS is used to optimize the satisfaction of the agent intentions. Moreover, intention consistency can be checked over the CPS structure. One of our next perspectives will consist to reinforce the proposed guidance service and temporal model checking techniques could be used in that sense.

References

1. Guivarch, V., Camps, V., Péninou, A.: Context awareness in ambient systems by an adaptive multi-agent approach. In: Paternò, F., de Ruyter, B., Markopoulos, P., Santoro, C., van Loenen, E., Luyten, K. (eds.) AmI 2012. LNCS, vol. 7683, pp. 129–144. Springer, Heidelberg (2012)
2. Olaru, A., Florea, A.M., El Fallah Seghrouchni, A.: A context-aware multi-agent system as a middleware for ambient intelligence. MONET **18**(3), 429–443 (2013)
3. Pi, N.S., Griol, D., Carbó, J., López, J.M.M.: Evaluation of agents interactions in a context-aware system. T. Comput. Collective Intell. **9**, 79–97 (2013)
4. Georgeff, M., Pell, B., Pollack, M.E., Tambe, M., Wooldridge, M.J.: The belief-desire-intention model of agency. In: Papadimitriou, C., Singh, M.P., Müller, J.P. (eds.) ATAL 1998. LNCS (LNAI), vol. 1555, pp. 1–10. Springer, Heidelberg (1999)
5. Chaouche, A.C., El Fallah Seghrouchni, A., Ilié, J.M., Saïdouni, D.E.: A higher-order agent model for ambient systems. Procedia Comput. Sci. **21**, 156–163 (2013)
6. Sardina, S., de Silva, L., Padgham, L.: Hierarchical planning in BDI agent programming languages: a formal approach. In: AAMAS '06, pp. 1001–1008 (2006)

7. Icard III, T.F., Pacuit, E., Shoham, Y.: Joint revision of beliefs and intention. In: Principles of Knowledge Representation and Reasoning, pp. 572–574 (2010)
8. Shapiro, S., Sardina, S., Thangarajah, J., Cavedon, L., Padgham, L.: Revising conflicting intention sets in BDI agents. In: AAMAS '12, pp. 1081–1088 (2012)
9. Hoek, W., Jamroga, W., Wooldridge, M.: Towards a theory of intention revision. Synthese 155(2), 265–290 (2007)
10. Meneguzzi, F., Zorzo, A.F., da Costa Móra, M., Luck, M.: Incorporating planning into BDI agents. Scalable Comput. Pract. Experience 8, 15–28 (2007)
11. Cohen, P.R., Levesque, H.J.: Intention is choice with commitment. Artif. Intell. 42(2–3), 213–261 (1990)
12. d'Inverno, M., Kinny, D., Luck, M., Wooldridge, M.: A formal specification of dmars. In: Rao, A., Singh, M.P., Wooldridge, M.J. (eds.) ATAL 1997. LNCS, vol. 1365, pp. 155–176. Springer, Heidelberg (1998)
13. Chaouche, A.C., El Fallah Seghrouchni, A., Ilié, J.M., Saïdouni, D.E.: A dynamical plan revising for ambient systems. Procedia Comput. Sci. 32, 37–44 (2014)
14. Brinksma, E. (ed.): ISO 8807, LOTOS - A Formal Description Technique Based on the Temporal Ordering of Observational Behaviour (1988)
15. Galván-Tejada, C.E., Garca-Vázquez, J.P., Garc-Ceja, E., Carrasco-Jiménez, J.C., Brena, R.F.: Evaluation of four classifiers as cost function for indoor location systems. Procedia Comput. Sci. 32, 453–460 (2014)
16. Gansemer, S., Grossmann, U., Hakobyan, S.: Rssi-based euclidean distance algorithm for indoor positioning adapted for the use in dynamically changing wlan environments and multi-level buildings. In: Indoor Positioning and Indoor Navigation (IPIN), pp. 1–6 (2010)
17. Olaru, A., Florea, A.M., El Fallah Seghrouchni, A.: Graphs and patterns for context-awareness. In: Novais, P., Preuveneers, D., Corchado, J.M. (eds.) ISAmI 2011. AISC, vol. 92, pp. 165–172. Springer, Heidelberg (2011)
18. Singh, D., Sardina, S., Padgham, L., James, G.: Integrating learning into a BDI agent for environments with changing dynamics. In: Toby Walsh, C.K., Sierra, C. (eds.) Proceedings of IJCAI'11, pp. 2525–2530 (2011)
19. Grant, J., Kraus, S., Perlis, D., Wooldridge, M.: Postulates for revising bdi structures. Synthese 175(1), 39–62 (2010)
20. Hindriks, K.: Twenty years of engineering multiagent systems. In: Keynote of the EMAS Workshop, Part of AAMAS'14 (2014)
21. Baljak, V., Benea, M.T., El Fallah Seghrouchni, A., Herpson, C., Honiden, S., Nguyen, T.T.N., Olaru, A., Shimizu, R., Tei, K., Toriumi, S.: S-claim: an agent-based programming language for Am I, a smart-room case study. Procedia Comput. Sci. 10, 30–37 (2012)
22. Bordini, R.H., Hübner, J.F., Vieira, R.: Jason and the golden fleece of agent-oriented programming. In: Bordini, R.H., Dastani, M., Dix, J., El Fallah Seghrouchni, A. (eds.) Multi-agent Programming: Languages, Platforms and Applications, pp. 3–37. Springer, New York (2005)
23. Dastani, M.: 2apl: a practical agent programming language. Auton. Agent. Multi-Agent Syst. 16(3), 214–248 (2008)
24. Lützenberger, M., Küster, T., Konnerth, T., Thiele, A., Masuch, N., Heßler, A., Keiser, J., Burkhardt, M., Kaiser, S., Albayrak, S.: Jiac v: A MAS framework for industrial applications. In: Proceedings of AAMAS '13, pp. 1189–1190 (2013)
25. Sudeikat, J., Braubach, L., Pokahr, A., Lamersdorf, W., Renz, W.: Validation of BDI agents. In: Bordini, R.H., Dastani, M., Dix, J., El Fallah Seghrouchni, A. (eds.) PROMAS 2006. LNCS (LNAI), vol. 4411, pp. 185–200. Springer, Heidelberg (2007)

26. Dennis, L.A., Fisher, M., Webster, M.P., Bordini, R.H.: Model checking agent programming languages. Autom. Softw. Engg. **19**(1), 5–63 (2012)
27. Dahmani, D., Ilié, J.-M., Boukala, M.: Time recursive petri nets. In: Jensen, K., van der Aalst, W.M.P., Billington, J. (eds.) ToPNoC I. LNCS, vol. 5100, pp. 104–118. Springer, Heidelberg (2008)
28. Saïdouni, D.E., Bouneb, M., Ilié, J.M.: Maximality semantic for recursive petri nets. In: ECMS, pp. 544–550 (2013)
29. Kouah, S., Saïdouni, D.E., Ilié, J.M.: Synchronized petri net: a formal specification model for multi agent systems. J. Softw. **8**(3), 587–602 (2013)
30. Ziparo, V.A., Iocchi, L., Nardi, D., Palamara, P.F., Costelha, H.: Petri net plans: a formal model for representation and execution of multi-robot plans. In Padgham, L., Parkes, D.C., Mller, J.P., Parsons, S. (eds.) AAMAS (1), IFAAMAS, pp. 79–86 (2008)
31. Mazouzi, H., El Fallah-Seghrouchni, A., Haddad, S.: Open protocol design for complex interactions in multi-agent systems. In: AAMAS'02, pp. 517–526 (2002)

An Ontological Consensus Augmented Framework for Collaborative Business Process Formulation

Hanh H. Hoang[1(⊠)], Trung V. Nguyen[1], and Vu Minh Hoang[2]

[1] Hue University, 3 Le Loi Street, Hue, Vietnam
{hhhanh,nvtrung}@hueuni.edu.vn
[2] Vietnam Posts and Telecommunications, Hue, Vietnam
minhvu@hue.vnn.vn

Abstract. Dynamic cross-enterprise collaboration through business process management is one of challenges on the business-to-business integration (B2Bi) research nowadays. Semantic Web-based approaches for BPM have been foreseen as a promising solution with taking advantages of Semantic Web technologies such as ontologies, semantic web services. With the support of Semantic Web technologies, the gap between business and information technology (IT) communities has been reduced in order to tackle the challenge. Together with the development of web services and its augmented extension with Semantic Web—Semantic Web Services (SWS), recent approaches in B2Bi research focus on their business process integration between enterprises using SWS for more appropriate service discovery. Taking into account the challenge of dynamically forming collaborative business processes for B2Bi form business view into their execution, in this article, we introduce an ontological approach for forming collaboration of business processes agilely within our BizKB framework; moreover, we use the Consensus theory in distributed knowledge processing in our novel method for service discovery.

Keywords: Business process modelling · BPM · Collaborative business processes · Business-to-business · B2B · B2B integration · Semantic BPM · Ontology · Consensus theory · Semantic web services

1 Motivation

Business-to-business integration (B2Bi) or so-called cross-enterprise collaboration in several contexts is one of priority strategies of the e-business research to improve enterprise excellences [1]. It requires exchanging and sharing business processes between business partners such as customers, suppliers, and distributors. One of the most important challenges in integrating or collaborating between companies in the e-business environment is how to collaborate business processes automatically, more accurately, flexibly and effectively. The success of the integration between businesses requires forming and managing collaborative business processes to achieve business goals. This process includes from high level of business process collaboration which is closer to business community, to lower levels in view of the execution of formed

© Springer-Verlag Berlin Heidelberg 2014
R. Kowalczyk and N.T. Nguyen (Eds.): TCCI XVI, LNCS 8780, pp. 170–189, 2014.
DOI: 10.1007/978-3-662-44871-7_7

collaborative processes through web services which are more IT relevance. Therefore, business process modelling has been considered as a mediation for the fusion of IT and business community for the better of cross-enterprise collaboration.

Semantic business process management (SBPM) emerged as a promising solution to bridge the gap between businesses and information technology field with the approach to perform business actions which are supported by the information technology with perspective of business process rather than technical perspective [2]. Managing businesses processes shall include methods, techniques and tools to support in designing and constructing rules, managing and analysing businesses operations. This is crucial in this new emergent data-intensive environment for e-business today as every business now has to see that their business processes are very precious data of their own [3]. However, handling of the BPM automatically in integrating business processes among enterprises is still low due to the interaction between the business process collaboration's semantics. To solve this problem, many researchers have recently proposed solutions in apply artificial intelligence in managing the processes of the collaboration between enterprises discussed in [4].

In our previous work [5], we proposed an approach called Ontological HTN (O-HTN) based on HTN Planning [6] and Web Service Modeling Ontology (WSMO) for forming collaborative business processes (CBP) dynamically in the problem of the cross-enterprise collaboration. The research results CBP formed with help of O-HTN and attached services profiles. The next phase is to discover the appropriate web services to match with service profiles kept in ontologies. In this article, while mentioning our recent research work in the BizKB framework, we emphasise on introducing a novel approach using consensus theory which is originated from solving conflict of data versions [7].

With these motivations, the article is structured as follows: BizKB Framework [4] is briefly described in the following section. In this section, we also introduce HTN-augmented in support for an effective decomposition of process from its high level in context of forming collaborative process in phases for the business collaboration. In Sects. 3 and 4, we focus our research on the discovery of web services for execution level using consensus theory. We propose algorithm for matching in view of its usage for web services discovery. The article is concluded with a sketch of future work.

2 BizKB Approach for B2Bi

The ultimate goal of our BizKB approach is to build a platform for BP discovery and integration based-on Semantic Web technologies, which supports the process of cross-enterprise collaboration. Many initiatives restrict the range of standards they deal with for political, practical or technical reasons. For companies exposed to different national, industry or enterprise-specific standards – as is practically every business if all of its communications are addressed – this approach is clearly of low practical value. A universally usable methodology will avoid the predefinition of a range of manageable standards [8].

2.1 BizKB Framework

As depicted in Fig. 1, the overall conceptual architecture of the BizKB framework consists of two main parts: the BizKB and the Process Formulator. The output of BizKB framework is CBP with Semantic Web Services profiles attached to the CBP. BizKB is the heart of the BizKB Framework which contains the knowledge of the businesses in the form of Business Process Modeling Ontology (BPMO)-based collaborative business processes with different levels of the abstraction [4].

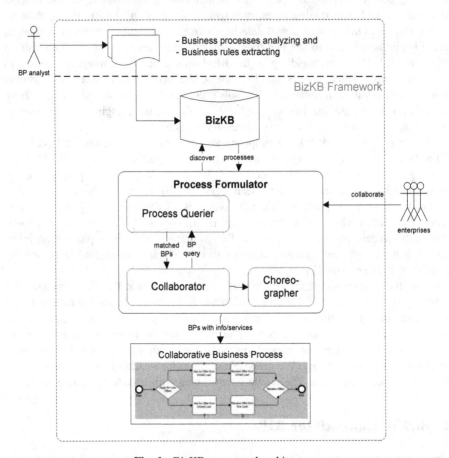

Fig. 1. BizKB conceptual architecture

In order to formulate these BPMO-based processes to store in the BizKB, the BP analysts are required as an important human factor of the system. Based on the analysis on the BPs, the found CBP patterns, level of the abstraction and associate business rules are also extracted and realised.

As described in Fig. 1, extracted artifacts of BPs are modelled using BPMO according to specific domains and kept in the BizKB. This repository is considered as the process feeder for the later stage of the CBP pattern discovery and CBPs formulation. The interactive part of the BizKB framework (Fig. 1) is the *Process Formulator* component which consists of two main subparts – *Process Querier* and the *Collaborator*. These parts are interacted by the demanding enterprise to find out the appropriate CBP patterns to form a collaborative business process with the help of the third subpart – *Choreographer*.

2.2 BizKB Ontology for Collaborative Business Processes

2.2.1 BizKB Ontology - BO

Current approaches to BPM suffer from a lack of automation that would support a smooth transition between the business world and the IT world and do not fulfil the task of the automation of BP integration in the B2B integration context [8]. According to the analysis in [1], the current BPM standards for B2B integration are lack of semantics that cannot coordinate the automated and dynamic B2B integration process which assumes that the partners are unknown. SBPM that is, the combination of Semantic Web and Semantic Web Services technologies with BPM, has been proposed as a solution for overcoming these problems [8].

In our approach, we use BPMO as the modelling notation to represent the CBPs in a semantics-enriched manner. BPMO is based on WSMO which is an ontology language for Semantic Web Services. WSMO is a comprehensive framework for semantically enabled Service-Oriented Architecture (SOA) technology. It defines semantic description models for four top level notions along with respective reasoning support for managing these: ontologies, Web services, goals, and mediators. WSMO appears to be a suitable extension to BPM in order to overcome the missing paces in current BPM standards for B2B integration [8]. Furthermore, with this modelling methodology, we can tackle the difficulties as follows (Fig. 2):

– Ontologies allow to model processes and data as shared conceptual model; their formalization allows semantically enhanced information processing;
– Semantic annotations of Web services allow to precisely detect and automatically executed suitable business functionalities as well as to maintain them;

```
namespace { _"http://www.ip-super.org/ontologies/BPMO/bpmo-1-4-20080109#",
       wsmostudio _"http://www.wsmostudio.org#" }

ontology instanceOntology_Procurement
     nonFunctionalProperties
           wsmostudio#version hasValue "0.7.3"
     endNonFunctionalProperties

     importsOntology
           _"http://www.ip-super.org/ontologies/BPMO/bpmo-1-4-20080109#"

instance ReceiveMessageEvent_1223890337875_1300073394 memberOf
ReceiveMessageEvent
       hasHomeProcess hasValue Process_1223879899796_1863342747
       hasName hasValue "Receive Message"
       messageFrom hasValue SendMessageEvent_1223890337890_641380045
```

Fig. 2. Fragment example of the BPMO-based procurement process

– Goals allow to specify processes and tasks on the problem layer for which suitable Web services can be detected dynamically at execution time;
– Mediators allow handling potentially occurring mismatches on the data and the process level, therewith enabling semantically stable interchange within and across enterprises if this is not given in advance.

2.2.2 BO for CBP

From B2B collaboration phases, a comprehensive list of CBP tasks can be modelled in BO. First, the sequences and hierarchies of granular tasks were synthesised into the three B2B collaboration phases.

Tasks. Business processes are a set of ordered compound or primitive tasks. In BO, both compound tasks and primitive tasks and they both are modelled as tasks. The two tasks may be differentiated by their *hasMethod* property. Compound tasks have one or more *hasMethod* property since they can be decomposed; not primitive tasks. Figure 3 shows the "Buy" compound task and its properties (i.e. *hasMethod, hasActor, hasParent*).

```
relationInstance hasMethod(_"http://www.owl-
ontologies.com/Ontology1231158528.owl#B-Buy", _"http://www.owl-
ontologies.com/Ontology1231158528.owl#MethodBuy.A")
relationInstance hasActor(_"http://www.owl-
ontologies.com/Ontology1231158528.owl#B-Buy", BuyerActor)

relationInstance hasParent(_"http://www.owl-
ontologies.com/Ontology1231158528.owl#B-Buy", B2BCollaboration)
```

Fig. 3. Simple task description in BO

Methods. A single compound task may have more than one method to decompose into primitive tasks. Each method has a prescription for how to decompose some task into a set of subtasks, with different restrictions that must be satisfied in order for method to be applicable and also various constraints of the subtask and relationship among them (Fig. 4).

In short, BO contains a set of ordered compound or primitive task and methods. Compound tasks have one more *"hasMethod"* property since they can be decomposed into primitive tasks that can be performed directly using O-HTN. Each method has a prescription for how to decompose some task into a set of subtasks, with different

```
relationInstance hasSubtasks(_"http://www.owl-
ontologies.com/Ontology1231158528.owl#MethodBuy.A", _"http://www.owl-
ontologies.com/Ontology1231158528.owl#S-Setup")

relationInstance hasSubtasks(_"http://www.owl-
ontologies.com/Ontology1231158528.owl#MethodBuy.A", _"http://www.owl-
ontologies.com/Ontology1231158528.owl#A-Action")
```

Fig. 4. A local method definition with embedded criteria matching, and control flows of subtasks

restrictions that must be satisfied in order for method to be applicable and also various constraints of the subtask and relationship among them [5].

2.3   O-HTN Augmented BizKB

2.3.1   HTN in Brief

The nature of dynamic business process formulation greatly resembles HTN planning from the field of artificial intelligence (AI) planning [6]. In HTN planning, a goal to a problem is realised via a plan of simple steps generated by the dynamic decomposition of a hierarchical network of compound tasks into sub-tasks in a domain. The lowest level task is a primitive task. To decompose and chain task, the HTN planning algorithm matches the constraints with the criteria of the appropriate method.

For illustration, consider two methods of travel planning for the compound task *travel(x,y)* (Fig. 5). The choice whether to travel by taxi or by air depends on the distance between x and y. If the distance (i.e. the constraint) is large, *travel(x,y)* will be decomposed into sub-tasks via the method "travel by air"; if the distance is short, the *travel(x,y)* task will be decomposed into sub-tasks "travel by taxi". All tasks are represented in a network of parent-child relationships.

After the HTN planning algorithm traverses through the HTN recursively decomposing tasks according to the matching methods, a result (or plan) is generated for "travelling from University of Maryland (UMD) to Massachusetts Institute of Technology (MIT)" (Fig. 6). Thus, it can be seen that HTN planning decomposes and sequences tasks (e.g. *travel (UMD, MIT)*).

2.3.2   HTN and CSP Combination

Users require various types of information and constraints, and automatic service composition requires several rounds of planning, because of trial and error, or for flexibly coping with dynamic exceptions. Web service composition by a planner alone has limitations that apply to a more general and intelligent composition of services [9].

First, it is inefficient for autonomously finding a solution in planning, because it does not provide a suitable basis for dealing with the evaluation of planning results with constraints. Second, although it works well for task ordering in planning, it is not good for dealing with a user's various requests for information. As real-life problems

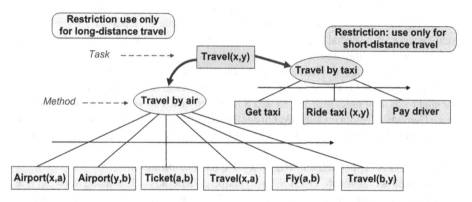

Fig. 5. A travel problem represented as a HTN

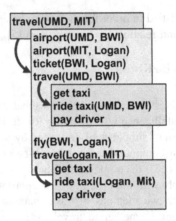

Fig. 6. A plan generated by the HTN algorithm

involve planning, scheduling, and executing, web service composition in real life requires not only planning information, but also additional information requests with constraints, which can be met by scheduling tasks jointly. A constraint satisfaction problem (CSP) formulation provides a strong basis for scheduling in a variety of real-life problems on the web. Third, it is weak regarding maintenance, because of the frequent invocation of services on the web. Although an Hierarchical Task Network (HTN) planner can invoke outside web services during planning, this causes severe restrictions and inefficiency, because service invocations in the planner are merged with the planning strategy [10, 11]. Combination of architecture planning and CSP help solve problems above. HTN and CSP combination is better than an HTN alone when problems involve scheduling plus other parameters.

We illustrate an example [10] as follows: A user, who lives in Aizu City in Japan, wants to go to South Carolina in the U.S.A for a vacation. If the user wants to go by train to the Narita international airport near Tokyo, there are three stages: by local train from Aizu to Koriyama to Tokyo, and by JR Express from Tokyo to Narita, from where a series of flights completes the journey to South Carolina.

Therefore, the user calls an agent to construct an itinerary to South Carolina. For this, the user provides basic information such as the departure date and location, and the arrival date and location. Suppose that the user wants to depart at 2:00 PM from Aizu because of a special business meeting. Therefore, he adds this new constraint to the basic input information.

Now, when the travel planner solves this problem, the solution may produce other internal spontaneous constraints temporarily. For instance, the planner should reserve a one-night stay in a hotel near Narita and a flight the next day when there is no flight to South Carolina at Narita on that day. On another occasion, the user may specify the arrival time in South Caroline as a constraint, which the planner will also need to accommodate.

2.3.3   O-HTN Approach for High-Level CBP Formulation

We proposed an approach, O-HTN for dynamic collaborative B2B using Web Service Modeling Ontology (WSMO) as the modelling foundation, WSMO is a flexible ontology language with dynamic reasoning features, supports execution based-on Web services as well. BO describe the hierarchical relationships between compound and primitive B2B collaboration tasks, and methods for task decompositions, and relevant planning criteria (e.g. cost, quantity ordered, type of collaboration) embedded in the methods. Different criteria input by the user result in different permutations of subtasks. Main reasons for the creation of O-HTN: O-HTN is feasible for dynamically creating CBP task sequences ideal for direct web services execution.

O-HTN Algorithm. Start with an initial high-level task and algorithm decomposes the task into subtasks, until primitive tasks are found that can be performed directly with web services. The O-HTN algorithm originates from [10, 11] and we have improved the algorithm which shown as follows.

```
Input: Task to be decomposed.
Output: Decomposed Tree, primitive actions.

Procedure HTNPlanning:
   Create empty tree
   /*decompose for three hierarchies of tasks for each
     collaboration phase*/
   Create three thread
   Each thread
      /*save Task when decompose in BO*/
      Decompose (nameTask, listMethod, listTask, criteria
      Return tree
End HTNPlanning

Procedure Decompose (nameTask, listMethod, listTask, criteria)
   Count number of methods in nameTask

   if there are no methods
      Mark task nameTask as primitive task for service execution
      Extract actor of task nameTask
      Write nameTask in tree
   else
      while there are methods for namTask not processed
         Select the next method nameMethod of nameTask
         Check supervised criteria of nameMethod with user criteria
         if supervised criteria matches user criteria
            Check number of control flows in nameMethod
         /* subtasks will be chained in control flows */
         while there are still control flows in nameMethod
            Read the outermost control flow cf
            Write the start of the cf in tree
            foreach subtask st in cf
              Decompose(st,"","",criteria)
              Write the end of cf in tree
         endwhile
         endif
      endwhile
   endif
End Decompose
```

2.3.4   O-HTN Advantages

Flexible collaboration at anywhere and anytime: Customers can access the system anywhere such as at office, at home, and at public Internet site, anytime.

Cost savings: Customers can save much money, mainly for collaboration between businesses. Therefore they can find potential partners anywhere without costly travel.

Flexibility: Customers can choose a partner that is most suitable to them with many diverse services.

Optimization: This framework can quickly assess customer need and then provide the collaborative models to meet the needs of customers.

Diversity: Many basic and specialized collaborative models can use the application O-HTN of collaboration between enterprises.

With the benefits listed above, the collaborative application O-HTN among businesses opens new opportunities in e-business environment. Customers may choose the appropriate partners dynamically. They also have many opportunities to contact with many new partners, save a lot of business resources, and access quality businesses (Fig. 7).

```
S-Setup
    Sequence
    D1-SearchForPotentialSuppliers
        Flow
        D1.1-ViaReferrals
            Sequence
            D1.1.1-ContactReferrals
            D1.1.2-ObtainPerspectivePartnersContactInfo
            SequenceEnd
        D1.2-SearchBizDirectory
        D1.3-SearchInternet
        D1.4-VisitingTradeShows
        D1.5-ProductCatalogInfo
            Sequence
            D1.5.2-ReceiveProductCatalogInfo
            SequenceEnd
        FlowEnd
    D4-PotentialPartnerListFound
    SequenceEnd
```

Fig. 7. Task tree is generate of algorithm

3   Consensus-Based Approach for Dynamic Service Discovery

3.1   Consensus Theory in Brief

Consensus methods were known in ancient Greece and were applied mainly in determining vote results. Along with the development of software methods we can see consensus methods can be applied into many applications fields, especially in solving conflicts and reconciling inconsistent data [12, 13]. Several results of consensus theory using for knowledge integration have been proposed in [12]. In this section, we introduce basic notations which are directly used in the problem of ontology integration.

By U we denote a finite set of objects representing possible values for a knowledge state. We also denote:

- $\pi_k(U)$ is the set of all k-element subsets (with repetitions) of set U. $k \in N$ (set of natural numbers).
- $\pi(U) = \bigcup_{k \in N} \pi_k(U)$ is the set of all nonempty subsets with repetitions of set U. An element in $\pi(U)$ is called as a conflict profile.

Definition 1 - Distance function. A distance function $d : U \times U \rightarrow [0, 1]$ is defined so that it has these following features:

(a) Nonnegative: $\forall x, y \in U : d(x, y) \geq 0$,
(b) Reflexive: $\forall x, y \in U : d(x, y) = 0 \Leftrightarrow x = y$,
(c) Symmetrical: $\forall x, y \in U : d(x, y) = d(y, x)$.

We call a space (U, d) which is defined in the above way is a distance space. With a $X \in \pi(U)$, we denote:

- $d(x, X) = \sum_{y \in X} d(x, y)$, with $x \in U$.
- $d_{t\_mean}(X) = \frac{\sum_{x, y \in X} d(x, y)}{k(k+1)}$, with $k = |X|$.
- $d_{min}(X) = \min\{d(x, X) : x \in U\}$.

Definition 2 - Consensus function. By a consensus function in space (U, d), we mean a function

$$C : \pi(U) \rightarrow 2^U.$$

For a conflict profile $X \in \pi(U)$, the set $C(X)$ is called the representation of X, and an element in X is called a consensus of profile X. X is a normal set (without repetitions).

Consensus functions need to satisfy some postulates [13] in order to elect "proper" representation(s) from a conflict profile. The mostly used consensus function are O_1-functions. The functions of this kind satisfy the so-called O_1 postulate [6]:

$$x \in C(X) \Rightarrow d(x, X) = \min_{y \in U} d(y, X), \text{with } \forall X \in \pi(U).$$

Definition 3 - Criteria for Consensus Susceptibility. Not from any conflict profile we can choose a consensus solution in general and O_1-consensus in specifically. We say that, profile X is susceptible to consensus in relation to postulate O_1 iff

$$d_{t\_mean}(X) \geq d_{min}(X).$$

3.2 Inconsistent Ontologies Integration at the Concept Level Using O1-function

Definition 4 – Ontology. Ontology is a quadruple $\langle C, I, R, Z \rangle$, where:

- C is a set of concepts (classes).
- I is a set of instances of concepts.
- R is a set of binary relations defined on C.
- Z is a set of axioms which are formulae of first-order logic and can be interpreted as integrity constraints or relationships between instances and concepts, and which cannot be expressed by the relations in set R, nor as relationships between relations included in R.

Definition 5 – The real world. Let A is a finite set of attributes. Each attribute $a \in A$ has a domain V_a. Let $V = \bigcup_{a \in A} V_a$, we call (A, V) as a real world.

A domain ontology that refers to the real world (A, V) is called as (A, V)-based.

Definition 6 – Structure of a concept. A concept in an (A, V)-based ontology is defined as a triple (c, A^c, V^c), where:

- c is the unique name of the concept,
- $A^c \subseteq A$ is a set of attributes describing the concept,
- $V^c = \bigcup_{a \in A^c} V_a$ is the domain of attributes. ($V^c \subseteq V$)

  Pair (A^c, V^c) is called the structure of concept c.

Definition 7 – Relations between attributes. Two attributes a, b in structure of a concept can have following relations:

- Equivalence: a is equivalence to b, denoted as $a \leftrightarrow b$, if a and b reflect the same feature for instances of the concept.
  For example, *occupation* $\leftrightarrow$ *job*.
- Generalization: a is more general than b, denoted as, $a \rightarrow b$, if information given by property a including information given by property b.
  For example: *dayOfBirth* $\rightarrow$ *age*.
- Contradiction: a is contradictory with b, denoted as $a \downarrow b$, if their domains are the same two-element set and values of them for the same instance are contradictory.
  For example: *isFree* $\downarrow$ *isLen*, where $V_{isFree} = V_{isLent} = \{true, false\}$ which can be used to describe instances in the Book concept whether its instances' property *isFree* changed to *isLent*.

Definition 8 - The ontology integration problem on the concept level. Let $O_1, O_2, \ldots, O_n (n \in N)$ are (A, V)-based ontologies. Let the same concept c belong to

O_i is (c, A^i, V^i), $i = 1, 2, \ldots, n$. From the profile $X = \{(A^i, V^i) : i = 1, 2, \ldots, n\}$ we have to determine the pair (A^*, V^*) which present the best structure for the concept c.

3.3   Postulates for Optimised integration (A^*, V^*).

Inspired by [7, 12, 13], we formulate the following postulates for determination of pair (A^*, V^*).

P1. For $a, b \in A = \bigcup_{i=1}^n A^i$ if $a \leftrightarrow b$ then all occurrences of a in all sets A^i may be replaced by attribute b or vice versa.

P2. If in any set A^i attributes a and b appear simultaneously and $a \rightarrow b$ then attribute b may be removed.

P3. For $a, b \in A = \bigcup_{i=1}^n A^i, a \downarrow b$, all occurrences of a in all sets A^i may be replaced by attribute b ore vice versa.

P4. Occurrence of an attribute in set A^* should be dependent only on the appearances of this attributes in sets A^i.

P5. An attribute a appears in set A^* if it appears in at least half of sets A^i.

P6. Set A^* is equal to A after applying postulates P1–P3.

P7. For each attribute $a \in A^*$, its domain V_a^* is determined so that:

$$d_a\left(V_a^*, X_a\right) = \min\{d(V_a, \ X_a) | V_a \in U_a\}$$

where:

– X_a is conflict profile which is formulated from domains $V_a^i, i = 1, \ldots, n$.
– U_a is universe set, contains possible values for V_a.

Postulates P1–P6 are adapted to ones in [13]. We propose the P7 postulate to gain the result of consensus theory. More specifically, we use the O_1 function to determine the optimal domain for the attribute $a \in A^*$. It is important to formulate distance space (U, d) for using O_1 function to find the consensus domain. The important issue is about appropriately defining space distance (U, d) to compute the optimised solution for the ranges of properties in integration set.

The remain part of this paper, we are going to describe the way how to formulate the space distance and integration algorithm according to these postulates.

3.4   Distance Function Between Two Concepts

There are several ways to measure the similarity between concepts in an ontology. In this article, we use idea of [14]. According to these authors, we allocate weight values to the edges between concepts:

$$w(parent, child) = 1 + \frac{1}{2^{depth(child)}}$$

where, $depth(child)$ present the depth of concept $child$ from the root concept in ontology hierarchy.

The so-called semantic distance between concepts can be determined using the following algorithm [14]:

Input: Concepts c_1, c_2 in ontology hierarchy.
Output: Semantic distance between c_1, c_2, denoted as $Sem\_Dis(c_1, c_2)$

Procedure Sem_Disc:
```
    if (c₁,c₂ is the same concept)
        Sem_Disc(c₁,c₂) = 0;
    else if (there exists the direct path relation between
                c₁and c₂)
        Sem_Disc(c₁,c₂) = w(c₁,c₂);
    else if (there exists the indirect path relations
                between c₁and c₂)
        Determine shortestPath(c₁,c₂) is the shortest path from
        c₁ to c₂ in the ontology hierarchy;
        Sem_Disc(c₁,c₂) = Σ(cᵢ,cⱼ) ∈ shortestPath(c₁,c₂) w(cᵢ, cⱼ)
    else
        Determine cpp = the nearest common parent concept of
        the two concepts c₁,c₂;
        Sem_Disc(c₁,c₂) = min{Sem_Disc(c₁,cpp)} + min { Sem_Disc(c₂,cpp)}
    endif
End Sem_Disc
```

We can see clearly that, the *Sem_Disc* function is not normalized, i.e. its values may be out of $[0, 1]$. We can normalise it and formulate a distance space (U, d) from the ontology hierarchy like this:

- U: the set of concepts in the ontology hierarchy.
- $d: U \times U \rightarrow [0, 1]$
  $$d(c_1, c_2) \mapsto 1 - \frac{1}{Sem\_Disc(c_1,c_2)+1}.$$

3.5   Optimal Integration Based on the Consensus

Based on postulates that are presented in Sect. 3.3, we propose an algorithm for determining integration structure for concept c from element ontologies $O_1, O_2, \ldots, O_n$ as follows.

Input:
- Conflict profile $X = \{(A^i, V^i), i = 1, \dots, n\}$, where (A^i, V^i) is structure of concept c in ontology O_i.
- Ontology hierarchy $O_{REF-TREE}$ that use for reference. $C_{REF-TREE}$ is the set of concepts in ontology $O_{REF-TREE}$.
- Distance space (U, d) that formulate for ontology $O_{REF-TREE}$ as in Sect. 3.4. $(U = C_{REF-TREE})$.

Output: Pair (A^*, V^*) present the best structure of concept c.

Procedure *Optimal_Integration*:

```
(1) Set A* := ∪ⁿᵢ₌₁ Aⁱ
(2) foreach a, b ∈ A*
    if (a ↔ b) and (a does not occur in relationships with
    other attributes from A*)
       set A* := A* \ {a} ;
    endif
    if (a → b) and (b does not occur in relationships with
    other attributes from A*)
       set A* := A* \ {b};
    endif
    if (a ↓ b) and (b does not occur in relationships with
    other attributes from A*)
       set A* := A* \ {a};
    endif
endfor

(3) foreach a ∈ A*
if (the number of occurrences of a in pairs {Aⁱ, Vⁱ} is
smaller than n/2)
  set A* = A* \ {a};
  else
    Set Xₐ = {V₁, V₂, ..., Vₖ} where Vⱼ is the domain of
    attribute a in pairs (Aⁱ, Vⁱ) and Vⱼ ∈ C_REF-TREE, j = 1,..,k,
    i = 1, ..., n;
    if (Xₐ is susceptible to consensus in relation to
    postulate to O₁)
       Determine Vₐ* as ₀₁ consensus in distance space
       (U, d): dₐ(Vₐ*, Xₐ) = min{d(Vₐ, Xₐ)| Vₐ ∈ U}
       Set Vₐ* as domain of attribute a in A*;
    else
       set A* := A* \ {a};
    endif
  endif
endfor

(4) foreach a from A*
    add to A* attribute b if there is relationship a ↔ b
    or a → b;
endfor
```

End *Optimal_Integration*

We have remarks for this algorithm:

- The complexity of the algorithm is $O(m^3)$ where $m = \left|\bigcup_{i=1}^{n} A^i\right|$ (m is the number of different attributes from sets $A^i, i = 1..n$).
- The algorithm only show way to determine optimal attributes for ones has domain in the ontology hierarchy $O_{REF-TREE}$. For attributes which have other kinds of domain (such as range of numbers) we can still use consensus theory to determine optimal domain.
- The algorithm determines consensus structure for concept c in both component: attributes and their domains.
- We can make some modifications in step (3) of algorithm to get some interesting results:

  - Get V_a^* from X_a rather than from U, i.e.: $d_a(V_a^*, X_a) = \min\{d(V_a, X_a) | V_a \in X_a\}$. We can use this condition in cases that $C_{REF-TREE}$ contains a large of concepts so that the algorithm has less time to run.
  - If X_a is not susceptible in relation to O_1, we can choose V_a^* as the common parent concept of concepts in X_a rather than remove attribute a from A^*.

3.6 Consensus-Based Collaborative Service Discovery

The ontology matching process is the result of alignments made by the semantic distance calculation. In order to discover the appropriate semantic web services, we use above algorithms for discovering the appropriate semantic web services from different partners which presented as profiles. We call our approach is "collaborative service discovery" because of searching the services from different partners with respect to their contexts and domain, and finding out a "consensus optimal" solution for appropriate services.

Input:
- Profile $X = \{(A^i, V^i), i = 1, ..., n\}$, where (A^i, V^i) is structure of concept c in ontology O_i of the corresponding semantic web services;
- Ontology hierarchy $O_{BizKB} = O_{REF-TREE}$ that use for reference in this case is the from BizKB.

Output: Structure of the concept c.

Procedure SWS_Discovery:
    Finding Optimal_Integration (X, O_{BizKB})
End SWS_Discovery

The result of the procedure will have a structure of the service profiles for the formed collaborative business process of previous step. This is one of the most important steps in BizKB framework for a dynamic approach from automatic decomposition of high-level process into collaborative processed attached profiles.

4 Collaborative Service Discovery for BizKB Framework

We have proposed an approach, O-HTN for dynamic collaborative B2B using Web Service Modeling Ontology (WSMO) as the modelling foundation [5]. The main reason for the creation of O-HTN-based BizKB is: O-HTN approach is feasible for dynamically creating CBP task sequences which is significant for a high level tasks composition for formed CBP. Next is about finding the appropriate web services to fill in the service profiles and this will be done by an ontology matching process.

In a collaborative environment, services are composed from different domains and contexts. Therefore, we can be challenged by the inconsistency of different knowledge-based domains. In our approach, we use the consensus methodology [12] to solve this problem.

4.1 Collaborative BizKB with O-HTN and Consensus Service Discovery Component

Applying these algorithms together with O-HTN in our approach, we structure the new process for CBP formulation and service discovery for BizKB framework as described in Fig. 8.

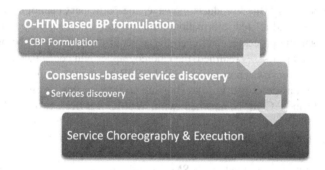

Fig. 8. A collaborative Biz-KB process with support of O-HTN and consensus service discovery mechanisms

Collaborative business processes are dynamically formed by using O-HTN algorithm in a flexible task decomposition process. Formed CBP with service profiles will then hand-in the services discovery phase for finding appropriate services.

The consensus service discovery based on newly invented ontology matching process helps to find out requested services in a better solution in collaborative service partner providers. The discovered services will be then transferred into the choreography mechanism for forming executable business processes.

4.2 Process Formulator Workflow

The O-HTN based Architecture for the Process Formulator is described in Fig. 9. The Process Querier helps to find the appropriate process patterns at a certain abstraction

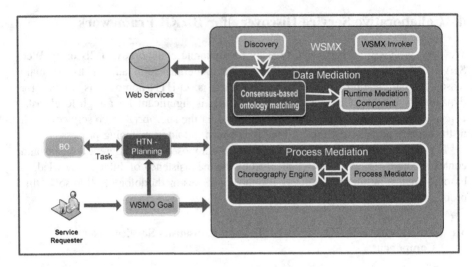

Fig. 9. The O-HTN-based process formulator architecture

level. Due to the enterprise's discovery into the BizKB, the detailed level will be matched to the need. For example, in the Order Management process, one participant wants to identify the process of "Buy" products, however the participant cannot clearly identify parts of the process and related information, the Process Querier can help to identify the basic patterns, sample processes, and even the generalization levels of the needed process. After matched processes returned, the Choreographer will coordinate to finalize the output collaborative business process to fulfil the B2B integration demand. Here, we use O-HTN Algorithm as described in following sub-section for this phase.

The new formed CBP is attached with WSMO services profiles for specific Semantic Web Services. This process is serialized using WSMO standard which conforms the unification of the framework's BPMO standard (which is based on WSMO) and benefits from semantic web services' advantages.

User's request is presented in WSMO ontologies and a WSMO Goal. Every Semantic Web service has a specific choreography that describes the way in which the user should interact with it. This choreography describes semantically the control and data flow of messages the Web Service can exchange. In cases where the choreography of the user and the choreography of the Web Service do not match, process mediation is required. The process Mediation component is WSMX is responsible for resolving mismatches between the choreographies of the user and web service.

If there is no single web service that satisfies the request then the request will be offered to the planner. The planner then tries to combine existing Semantic Web services and generate the process model. In the proposed framework, the process generator is based on HTN-planning. The process generator to tackle the problems of heterogeneous ontologies and choreography uses discovery component of WSMX. Thus via this component, the process generator will be able to discover the appropriate Semantic Web services for dynamic cross-enterprise collaboration. Finally the process model will be offered to the WSMX for its execution. The stages for execution of Web services as a process model are like as single web services.

5 Related Work

Since the failure of the non-semantic approaches as mentioned above, research efforts have been emerged from the motivation of knowledge management and applying Semantic Web technologies into BPM researches to bring the administrative side and IT side together.

SUPER [11] addresses the ever enduring need of new weaponry in struggle for survival in optimistic business environment where profit margins dramatically drop while competitiveness reaches the new sky high limits. The major objective of the SUPER project is to raise BPM to the business level, where it belongs, from the IT level where it mostly resides now. This objective requires that BPM is accessible at the level of semantics of business experts. SUPER's approach has tried to transform existing BPMN and BPEL standards into a semantics-enriched form, respectively called sBPMN and sBPEL in the attempt to realize their goals.

In the same line, the SemBiz project (http://www.sembiz.org/) aims at bridging the gap between the business level perspective and the technical implementation level in Business Process Management (BPM) by semantic descriptions of business processes along with respective tool support. This approach used WSMO as a basis for defining an exhaustive semantic description framework for business processes. On basis of this, novel functionalities for BPM on the business level can be supported by inference-based techniques that work on semantic process descriptions.

Haller in [16] extended the *multi metamodel process ontology* (m3po) introduced with concepts for a full formalisation of the meta-model of XPDL. In the context of their approach, to deal with collaborative processes (choreographies) these internal workflow models are aligned to the external behaviour advertised through web services interfaces. The *m3po* ontology presented explicitly models the complete semantics of XPDL. The integrated *m3po* is used as shared representation to perform the integration. The advantage of this approach is that authors use a web ontology language to formalise proposed model into linked data with established business document standards.

One of recent efforts in cross-enterprise collaboration research is Genesis approach based on its ontology called Business-OWL (BOWL) [11]. The core of the approach is about BOWL that is a hierarchical task networking (HTN) [10] modelled in OWL describing the hierarchical relations between tasks of collaborative business processes consist of compound tasks, primitive tasks and task decomposition methods. For a suitable solution for current dynamic e-commerce today. Therefore, the knowledge described by HTN needs to be modelled in forms of OWL ontologies proposed in this approach. Following this research direction brings new opportunities, new prospects and useful tools for e-business and B2B integration especially. The effort follows this line is Jung's work [2] which focus on basic problems of applying ontology aligning for business process integration.

Consensus theory introduced in a research line [7, 12, 13] is becoming a power method for knowledge integration of distributed systems. This methodology has become a powerful tool for solving conflicts and inconsistency in ontological systems as data and information from heterogeneous sources. Therefore, in the context of ontology matching problems, we have used this theory for finding appropriate services for formed collaborative business processes in BizKB framework.

6 Conclusion

Business Process Management's efforts are to bring business and IT communities together to solve the problem of the integration and collaboration cross-enterprises. BPM's approach is to execute business tasks with the IT support in the business expert perspective rather than technical one. BPM is challenged with issues of the automation of dynamic business collaboration and the integration/collaboration of business process in B2Bi. In this article, we have introduced a framework which is a fusion of Semantic Web technologies, AI planning and Consensus theory for solving the challenge of dynamic B2B integration: formulating collaborative business process with understanding the semantics in a specific domain application and 'on the fly'.

Also in this article, we emphasise on a novel approach to formulate distance space from a similarity measure of concepts and use it for develop an algorithm determining integration of ontologies which are conflict on concept level. The article also showed that consensus theory is an appropriate and effective way for ontology integration problem in view of its usage for semantic web service discovery. For the next steps, we plan to fully developed this framework for more several business cases.

Acknowledgement. This work was generously sponsored by Vietnam's National Foundation for Science and Technology Development (NAFOSTED) in the framework of research grant "Application of Semantic Web concepts for Collaborative Business Process Management" - 102.02-2010.14.

References

1. Hoang, H.H, Jung, J.J., Tran, C.P.: Ontology-based approaches for cross-enterprise collaboration: a literature review on semantic business process management. Enterp. Inf. Syst. (2013). doi:10.1080/17517575.2013.767382 (Taylor and Francis)
2. Jung, J.J.: Semantic business process integration based on ontology alignment. Expert Syst. Appl. **36**(8), 11013–11020 (2009). (Elsevier)
3. Vossen, G.: Big data as the new enabler in business and other intelligence. Vietnam J. Comput. Sci. **1**(1), 3–14 (2013). (Springer)
4. Hoang, H.H., Le, T.M.: BizKB: a conceptual framework for dynamic cross-enterprise collaboration. In: Nguyen, N.T., Kowalczyk, R., Chen, S.-M. (eds.) ICCCI 2009. LNCS, vol. 5796, pp. 401–412. Springer, Heidelberg (2009)
5. Hoang, V.M., Hoang, H.H.: An ontological approach for dynamic cross-enterprise collaboration. In: WAINA-2012, IEEE Computer Society, pp. 1355–1360. IEEE (2012)
6. Erol, K., Hendler, J., Nau, D. S.: HTN planning: complexity and expressivity. In: Proceedings of the National Conference on Artificial Intelligence, pp. 1123–1123. Wiley (1995)
7. Nguyen, N.T.: Consensus systems for conflict solving in distributed systems. Inf. Sci. **147** (1–4), 91–122 (2002). (Elsevier)

8. Pedrinaci, C., Domingue, J., Brelage, C., van Lessen, T., Karastoyanova, D., Leymann, F.: Semantic business process management: scaling up the management of business processes. In: Proceedings of 2008 IEEE International Conference on Semantic Computing, pp. 546–553. IEEE (2008)

9. Damodaran, S.: B2B integration over the Internet with XML: RosettaNet successes and challenges. In: Proceedings of the 13th international World Wide Web conference on Alternate track papers and posters, pp. 188–195. ACM (2004)

10. Paik, I., Maruyama, D., Huhns, M.N.: A framework for intelligent web services: combined HTN and CSP approach. In: Proceedings of IEEE International Conference on Web Services, pp. 959–962. IEEE (2006)

11. Ko, R.K.L., Lee, E.W., Lee, S.G.: Business-OWL (BOWL)—a Hierarchical task network ontology for dynamic business process decomposition and formulation. IEEE Trans. Serv. Comput. 5(2), 246–259 (2012). (IEEE)

12. Nguyen, N.T.: Using consensus methodology in processing inconsistency of knowledge. In: Last, M., Szczepaniak, P., Volkovich, Z., Kandel, A. (eds.) Advances in Web Intelligence and Data Mining, pp. 161–170. Springer, Heidelberg (2006)

13. Nguyen, N.T.: Processing inconsistency of knowledge in determining knowledge of a collective. Cybern. Syst. 40(8), 670–688 (2009). (Taylor and Francis)

14. Ge, J., Qiu, Y.: Concept similarity matching based on semantic distance. In: Proceedings of Fourth International Conference on Semantics, Knowledge and Grid 2008 (SKG'08), pp. 380–383. IEEE (2008)

15. Born, M., Drumm, C., Markovic, I., Weber, I.: SUPER - raising business process management back to the business level. ERCIM News 70, 43–44 (2007). (ERCIM)

16. Haller, A., Oren, E., Kotinurmi, P.: m3po: an ontology to relate choreographies to workflow models. In: Proceedings of IEEE International Conference on Services Computing (SCC'06), pp. 19–27. IEEE (2006)

17. Nurmilaakso, J.M.: EDI, XML and e-business frameworks: a survey. Comput. Ind. 59(4), 370–379 (2008). (Elsevier)

Author Index

Printed in the United States
By Bookmasters